# Packaged Pleasures

# Packaged Pleasures

*How Technology & Marketing Revolutionized Desire*

**Gary S. Cross**

**Robert N. Proctor**

THE UNIVERSITY OF CHICAGO PRESS　✳　CHICAGO & LONDON

**Gary S. Cross** is distinguished professor of modern history at
Pennsylvania State University and the author of many books, including
*All-Consuming Century: How Commercialism Won in Modern America*
and *The Playful Crowd: Pleasure Places in the Twentieth Century.*
**Robert N. Proctor** is professor of the history of science at Stanford
University and the author of many books, including *Racial Hygiene: Medicine
Under the Nazis* and *Value-Free Science? Purity and Power in Modern Knowledge.*

The University of Chicago Press, Chicago 60637
The University of Chicago Press, Ltd., London
© 2014 by The University of Chicago
All rights reserved. Published 2014.
Printed in the United States of America

23 22 21 20 19 18 17 16 15 14 1 2 3 4 5

ISBN-13: 978-0-226-12127-7 (cloth)
ISBN-13: 978-0-226-14738-3 (e-book)
DOI: 10.7208/chicago/9780226147383.001.0001

Library of Congress Cataloging-in-Publication Data
Cross, Gary S., author.
Packaged pleasures : how technology and marketing
revolutionized desire / Gary S. Cross and Robert N. Proctor.
pages ; cm
Includes bibliographical references and index.
ISBN 978-0-226-12127-7 (cloth : alk. paper) — ISBN 978-0-226-14738-3 (e-book)
1. Packaging—Technological innovations—Psychological aspects. 2. Packaging—
Technological innovations—Social aspects. 3. Marketing—Technological
innovations—Psychological aspects. 4. Marketing—Technological innovations—
Social aspects. 5. Consumer behavior. 6. Technological innovations—Psychological
aspects. 7. Technological innovations—Social aspects.
I. Proctor, Robert, 1954- author. II. Title.
T173.8.C767 2014
658.8′23019—dc23
2013049702

♾ This paper meets the requirements of
ANSI/NISO Z39.48-1992 (Permanence of Paper).

# Contents

**1**

# The Carrot and
# the Candy Bar

Our topic is a revolution—as significant as anything that has tossed the world over the past two hundred years. Toward the end of the nineteenth century, a host of often ignored technologies transformed human sensual experience, changing how we eat, drink, see, hear, and feel in ways we still benefit (and suffer) from today. Modern people learned how to capture and intensify sensuality, to preserve it, and to make it portable, durable, and accessible across great reaches of social class and physical space. Our vulnerability to such a transformation traces back hundreds of thousands of years, but the revolution itself did not take place until the end of the nineteenth century, following a series of technological changes altering our ability to compress, distribute, and commercialize a vast range of pleasures.

Strangely, historians have neglected this transformation. Indeed, behind this astonishing lapse lies a common myth—that there was an age of production that somehow gave rise to an age of consumption, with historians of the former exploring industrial technology, while historians of the latter stress the social and symbolic meaning of goods. This artificial division obscures how technologies of production have transformed what and how we actually consume. Technology does far more than just increase productivity or transform work, as historians of the Industrial Revolution so often emphasize. Industrial technology has also shaped how and how much we eat, what we wear and why, and how and what (and how much!) we hear and see. And myriad other aspects of how we experience daily life—or even how we long for escape from it.

Bound to such transformations is a profound disruption in modern life, a breakdown of the age-old tension between our bodily desires and the scarcity of opportunities for fulfillment. New technologies—from the rolling of cigarettes to the recording of sound—have intensified the gratification of desires but also rendered them far more easily satisfied, often to the point of grotesque excess. An obvious example is the mechanized packaging of highly sugared foods, which began over a century ago and has led to a health and moral crisis today. Lots of media attention has focused on the irresponsibility of the food industry and the rise of recreational and workplace sedentism—but there are other ways to look at this.

It should be obvious that technology has transformed how people eat, especially with regard to the *ease* and *speed* with which it is now possible to ingest calories. Roots of such transformations go very deep: the Neolithic revolution ten-plus thousand years ago brought with it new methods of regularizing the growing of food and the world's first possibility of elite obesity. The packaged pleasure revolution in the nineteenth century, however, made such excess possible for much larger numbers of "consumers"—a word only rarely used prior to that time. Industrial food processors learned how to pack fat, sugar, and salt into concentrated and attractive portions, and to manufacture these cheaply and in packages that could be widely distributed. Foods that were once luxuries thus became seductively commonplace. This is the first thing we need to understand.

We also need to appreciate that responsibility for the excesses of today's consumers cannot be laid entirely at the doors of modern technology and the corporations that benefit from it. We cannot blame the food industry alone. No one is forced to eat at McDonald's; people choose Big Macs with fries because they satisfy with convenience and affordability, just as people decide to turn on their iPods rather than listen to nature or go to a concert. But why would we make such a choice—and is it entirely a "free choice"? This brings us to a second crucial point: humans have evolved to seek high-energy foods because in prehistoric conditions of scarcity, eating such foods greatly improved their ancestors' chances of survival. This has limited, but not entirely eliminated, our capacity to resist these foods when they no longer are scarce. And if we today crave sugar and fat and salt, that is partly because these longings must have once promoted survival, deep in the pre-Paleolithic and Paleolithic. Our taste buds respond gleefully to sugars because we are descended from herbivores and especially frugivores for whom sweet-tasting plants and fruits were neuro-marked as edible and nutritious. Poisonous plants were more often bitter-tasting. Pleasure at least in this sensory sense was often a clue to what might help one survive.

But here again is the rub. Thanks to modern industrialism, high-calorie foods once rare are now cheap and plentiful. Industrial technology has overwhelmed and undercut whatever balance may have existed between the biological needs of humans and natural scarcity. We tend to crave those foods that before modern times were rare; cravings for fat and sugar were no threat to health; indeed, they improved our chances of survival. Now, however, sugar, especially in its refined forms, is plentiful, and as a result makes us fat and otherwise unhealthy. And what is true for sugar is also true for animal fat. In our prehistoric past fat was scarce and valuable, accounting for only 2 to 4 percent of the flesh of deer, rabbits, and birds, and early humans correctly gorged whenever it was available. Today, though, factory-farmed beef can consist of 36 percent fat, and most of us expend practically no energy obtaining it. And still we gorge.[1]

And so the candy bar, a perfect example of the engineered pleasure, wins out over the carrot and even the apple. More sugar and seemingly more varied flavors are packed into the confection than the unprocessed fruit or vegetable. In this sense our craving for a Snickers bar is partly an expression of the chimp in us, insofar as we desire energy-packed foods with maximal sugars and fat. The concentration, the packaging, and the ease of access (including affordability) all make it possible—indeed enticingly easy—to ingest far more than we know is good for us. Our biological desires have become imperfect guides for good behavior: drives born in a world of scarcity do not necessarily lead to health and happiness in a world of plenty.[2]

But food is not the only domain where such tensions operate. Indeed, a broader historical optic reveals tensions in our response to the packaged provisioning of other sensations, and this broader perspective invites us to go beyond our current focus on food, as important as that may be.

As biological creatures we are naturally attracted to certain sights and sounds, even smells and motion, insofar as we have evolved in environments where such sensitivities helped our ancestors prevail over myriad threats to human existence. The body's perceptual organs are, in a sense, some of our oldest tools, and much of the pleasure we take in bright colors, combinations of particular shapes, and certain kinds of movement must be rooted in prehistoric needs to identify food, threats, or mates from a distance. Today we embrace the recreational counterparts, filling our domestic spaces with visual ornaments, fixed or in motion, reminding ourselves of landscapes, colors, or shapes that provoke recall or simulate absent or even impossible worlds.

What has changed, in other words, is our access to once-rare sensations, including sounds but especially imagery. The decorated caves of southern France, once rare and ritualized space, are now tourist attractions, accessible to all through electronic media. Changes in visual technology have made possible a virtual orgy of visual culture; a 2012 count estimated over 348,000,000,000 images on the Internet, with a growth rate of about 10,000 per second.[3] The mix and matrix of information transfer has changed accordingly: orality (and aurality) has been demoted to a certain extent, first with the rise of typography (printing) and then the published picture, and now the ubiquitous electronic image on screens of different sorts. "Seeing is believing" is an expression dating only from about 1800, signaling the surging primacy of the visual. Civilization itself celebrates the light, the visual sense, as the darkness of the night and the narrow street gradually give way to illuminated interiors, light after dark, and ever broader visual surveillance.[4]

Humans also have preferences for certain smells, of course, even if we are (far) less discriminating than most other mammals. Technologies of odor have never been developed as intensively as those of other senses, though we should not forget that for tens of thousands of years hunters have employed dogs—one of the oldest human "tools"—to do their smelling. Smell has also sometimes marked differences between tribes and classes, rationalizing the isolation of slaves or some other subject group. The wealthy are known to have defined themselves by their scents (the ancient Greeks used mint and thyme oils for this purpose), and fragrances have been used to ward off contagions. Some philosophers believed that the scent of incense could reach and please the gods; and of course the devil smelled foul—as did sin.[5]

Still, the olfactory sense lost much of its acuity in upright primates, and it is the rare philosopher who would base an epistemology on odor. Philosophers have always privileged sight over all other senses—which makes sense given how much of our brain is devoted to processing visual images (canine epistemology and agnotology would surely be quite different). Optico-centricity was further accentuated with the rise of novel ways of extending vision in the seventeenth century (microscopes, telescopes) and still more with the rise of photography and moving pictures. Industrial societies have continued to devalue scent, with some even trying to make the world smell-free. Pasteur's discovery of germs meant that foul air (think miasma) lost its role in carrying disease, but efforts to remove the germs that caused such odors (especially the sewage systems installed in cities in the nineteenth century) ended up mollifying much of the stink of large urban centers. Bodily perfuming has probably been around for as long as humans have been human, but much of recent history has involved a process of deodorizing, further reducing the value of the sensitive nose.

Modern people may well gorge on sight, but we certainly remain sound-sensitive and long for music, "the perfume of hearing" in the apt metaphor of Diane Ackerman. Music has always aroused a certain spiritual consciousness and may even have facilitated social bonding among early humans. Stringed and drum instruments date back only to about 5,500 years ago (in Mesopotamia), but unambiguous flutes date back to at least 40,000 years ago; the oldest known so far is made from vulture and swan bones found in southern Germany.[6] Singing, though, must be far older than whatever physical evidence we have for prehistoric music.[7]

There is arguably a certain industrial utility to music, insofar as "moving and singing together made collective tasks far more efficient" (so claims historian William McNeill). As a mnemonic aid, a song "hooks onto your subconscious and won't let go." Music carries emotion and preserves and transports feelings when passed from one person or generation to another—think of the "Star Spangled Banner" or "La Marseillaise." And music also marks social differences in stratified societies. In Europe by the eighteenth century, for example, people of rank had abandoned participation in the sounds and music of traditional communal festivals and spectacles. To distinguish themselves from the masses, the rich and powerful came to favor the orderly stylized sounds of chamber music—and even demanded that audiences keep silent during performances. One of the signal trends of this particular modernity is the withdrawal of elites from public festivals, creating space instead for their own exclusive music and dance to eliminate the unruly/unmanaged sounds of the street and work. Music helps forge social bonds, but it can also work to separate and to isolate, facilitating escape from community (think earbuds).[8]

We humans also of course crave motion and bodily contact, flexing our muscles in the manner of our ancestors exhilarating in the chase. And even if we no longer chase mammoth herds with spears, we recreate elements of this excitement in our many sports, testing strength against strength or speed against speed, forcing projectiles of one sort or another into some kind of target. Dance is an equally ancient expression of this thrill of movement, with records of ritual motion appearing already on cave and rock walls of early humans. The emotion-charged dance may be diminished in elite civilized life, but it clearly reappears in the physicality of amusement park throngs at the end of the nineteenth century, and more recently in the rhythmic motions of crowds at sporting events and rock concert moshing where strangers slam and grind into each other.[9]

Sensual pleasure is thus central to the "thick tapestry of rewards" of human evolutionary adaptation, rewards wired into the complex circuitry of the brain's pleasure centers.[10] Pursuit of pleasure (and avoidance of pain) was certainly not an evil in our distant past; indeed, it must have had obvious advantages in promoting evolutionary fitness. Along with other adaptive emotions (fear, surprise, and disgust, for example), pleasure and its pursuit must also have helped create capacities to bond socially—and perhaps even to use and to understand language. The joy that motivates babies to delight in rhythmic and consonant sounds, bright colors, friendly faces, and bouncing motion helps build brain connections essential for motor and cognitive maturity.

Of course the biological propensity to gorge cannot be new; that much we know from the relative constancy of the human genetic constitution over many millennia. We also know that efforts to augment or intensify sensual pleasure long predate industrial civilization. This should come as no surprise, given that, as already noted, our longings for rare delights of taste, sight, smell, sound, and motion are rooted in our prehistoric past. Humans—like wolves—have been bred to binge. But in the past, at least, nature's parsimony meant that gorging was generally rare and its impact on our bodies, psyches, and sociability limited.

This leads us again to a critical point: pleasure is born in its paucity and scarcity sustains it. And scarcity has been a fact of life for most of human history; in fact, it is very often a precondition for pleasure. Too much of any good can lead to boredom—that is as true for music or arcade games as for ice cream or opera. Most pleasures seem to require a context of relative scarcity. Amongst our prehistoric ancestors this was naturally enforced through the rarity of honey and the all-too-infrequent opportunity for the chase. Humans eventually developed the ability, however, to create and store surpluses of pleasure-giving goods, first by cooking and preserving foods and drinks and eventually by transforming even fleeting sensory experiences into reproducible and transmissible packets of pleasure. Think about candy bars, soda pop, and cigarettes, but also photography, phonography, and motion pictures—all of which emerged during the packaged pleasure revolution.

Of course, in certain respects the defeat of scarcity has a much older history, having to do with techniques of containerization. Prior to the Neolithic, circa ten thousand years ago, humans had little in the way of either technical means or social organization to store *any* kind of sensual surplus (though meats may have been stashed the way some nonhuman predators do). Farming and its associated technics changed this. After hundreds of thousands of years of scavenging and predation, people in this new era began to grow their own food—and then to save and preserve it in containers, especially in pots made from clay but also in bags made from skins or fibers from plants. Agriculture seems to have led to the world's first conspicuous inequalities in wealth, but also the first routine encounters with obesity and other sins of the flesh (drunkenness, for example). Of course the rich—the rulers and priests of ancient city-states and empires or the lords and abbots of religious centers in the Middle Ages—were able to satisfy sensual longings more often, and in some cases continually.

While Christianity was in part a reaction to this sensual indulgence, being originally a religion of the excluded slave and the appalled rich, medieval aristocrats returned to the ancient love of sweet and sour dishes, favoring roasted game (a throwback to the preagricultural era) and the absurd notion that torturing animals before killing them made for the tastiest meats. Medieval European nobility mixed sex, smell, and taste in their large midday meals and frequent evening banquets.[11] Christian church fathers banned perfumes and roses as Roman decadence, but treatments of this sort—along with passions for pungent flavors and scents—were revived with the Crusades and intimate contact with the Orient.[12]

Until recently, pursuit of pleasure on such an opulent scale was confined to those tiny minorities with regular access to the resources to contain and intensify nature. Since antiquity, in fact, the powerful have often been snobbish killjoys, trying to restrict what the poor were allowed to eat, wear, and enjoy. Sometimes this made economic (if invidious) sense—as when England's Edward III rationed the diet of servants during shortages that followed the Black Death. In the sixteenth century, French law prohibited the eating of fish and meat at the same meal in hopes of preserving scarce supplies. And given the low output of agriculture, there was a certain logic underlying the rationing of access to "luxuries." But the powerful sometimes seem to have relished denying pleasure to others. How else do we explain sumptuary laws that prohibited the commoner from wearing colorful and costly clothing reserved for aristocrats?[13]

Access to pleasure has long been an expression of privilege and power, but much can be made with little, and rarely has pleasurable display been totally suppressed in any culture. Think of the ceremonies surrounding seasonal festivals, especially the gathering of harvest surplus, when humans drenched themselves in the senses that seemed almost to ache for expression. Think of the Bacchanalia of the Greeks, the Saturnalia of the Romans, the Mardi Gras of medieval Europeans, or the orgies of feasting, dancing, music, and colorful costumes of any society whose everyday world of scarcity is forgotten in bingeing after harvest. Agriculture produced cycles of carnival and Lent, "a self-adjusting gastric equilibrium," in the words of one historian.[14]

Of course there are many examples of ancient philosophers and sages seeking to limit the hedonism of the privileged (and the festival culture of the poor). Certainly there are ancients who embraced the virtues of moderation, as in Aristotle's "golden mean" or Confucian ideals of restrained desire. Hebrew prophets, Puritans, Jesuits, and countless Asian ascetics likewise attempted to rein in the fêtes of the senses. Medieval authorities in Europe forbade the eating of meat on Wednesdays, Fridays, and numerous fast days that added up to more than 150 days a year. The classical ideal of moderation was revived, and the moral superiority of grain-based foods was defended. Gluttony was condemned along with lust. Pleasure was to be regulated even in the afterlife, insofar as the Christian heaven was not for pleasure but for self-improvement. These and other ascetic moralities arguably helped people cope with uncertain supplies, putting a brake also on the rapacious greed of the rich and powerful. Curbing of excess extended to all manner of "pleasures of the flesh," including those that, like sex, were not necessarily even scarce.[15]

Dance came under suspicion in this regard, especially in its ecstatic form. European explorers frowned on the gesticulations of "possessed natives" whom they encountered in Africa and the Americas in the sixteenth and seventeenth centuries. At the same time, European elites smothered social dancing in the towns and villages of their own societies. The reasons were many. Clergy demanded that their holy days and rituals be protected from defilement by the boisterous and even sacrilegious customs of the frolicking crowd; the rich also chose to withdraw from—and then suppressed—the emotional intensity of common people's celebrations, retiring instead to the confines of their private gatherings and sedate dances. The military also needed a new type of soldier and new ways of preparing men for war: the demand was no longer to fire up the emotions of soldiers to prepare them for hand-to-hand combat; the new need was to drill and discipline troops to march unflinching into musket and cannon fire, with individual fighters acting as precision components in a machine. The regular rhythms of the military march served this purpose better than the ecstatic dance.[16]

Even when people found ways of intensifying sensation (as in the distillation of alcoholic spirits), state and church authorities were often able to enforce limits, sometimes by harsh means. In London in the 1720s, authorities repressed the widespread and addictive use of gin (a juniper-flavored liquor). At the beginnings of the Industrial Revolution, just as unleashing desire was becoming respectable, philosophers such as Adam Smith and David Hume still mused about the need for personal restraint and moral sympathies.

By this time, and increasingly over the course of the nineteenth century, especially between about 1880 and 1910, these traditional calls for moderation and self-control were starting to face a new kind of challenge, thanks to new techniques of containerization and intensification that would culminate in the packaged pleasure revolution. New kinds of machines brought new sensations to ordinary people, producing goods that for the first time could be made quite cheap and easily storable and portable. Canned food defeated the seasons, extending the availability of fruits and vegetables to the entirety of the year. Candy bars purchased at any newsstand or convenience store replaced the rare encounter with the honeycomb or wild strawberry. And while our more immediate predecessors may have enjoyed a pipe of tobacco or a draft of warm beer, the deadly convenience of the cigarette and the refreshing coolness of the chilled beverage came within the grasp of the masses only toward the end of the nineteenth century. And this revolution in the range and intensity of sensation radically upset the traditional relationship between desire and scarcity.

A similar process occurred with other sensory delights. While early-nineteenth-century Americans and Europeans thrilled at the sight of painted dioramas and magic lantern shows, nothing compared to the spectacle of fast-paced police chases in the one-reel movies viewable after 1900. Opera was a privileged treat of the few in lavish public places, but imagine the revolution wrought by the 1904 hard wax cylinder phonograph, when Caruso could be called upon to sing in the family parlor whenever (and however often) one wanted. Daredevils in Vanuatu dove from high places holding vines long before bungee jumping became a fad; even so, there was nothing like the mass-market calibrated delivery of physical thrills before the roller coaster, popularized in the 1890s. We find something similar even with binge partying: while peoples had long celebrated surpluses in festivals, they typically did so only on those rare days designated by the authorities. By the end of the nineteenth century, however, festive pleasures of a more programmed sort had become widely available on demand in the modern commercial amusement park.

Especially important is how the packaged pleasure intensified (certain aspects of) human sensory experience. An extreme example is when opium, formerly chewed, smoked, or drunk as tea, was transformed through distillation into morphine and eventually heroin — and then injected directly into the bloodstream with the newly invented syringe in the 1850s. The creation of a wide variety of "tubes" like the syringe for delivering chemically purified, intense sensation was characteristic of much of this new technology — which we shall describe in terms of "tubularization." The cigarette is another fateful example: tobacco smoking was made cheap, convenient, and "mild" (i.e., deadly) with the advent of James Bonsack's automated cigarette rolling machine (in the 1880s) and new methods of curing tobacco. Bonsack's machine lowered the cost of manufacturing by an order of magnitude, and new methods of chemical processing (such as flue curing) allowed a milder, less alkaline smoke to be drawn deep into the lungs. A new mass-market consumer "good" was born, accompanied by mass addiction and mass death from maladies of the heart and lungs.

The "tubing" of tobacco into cigarettes was closely related to techniques used in packing and packaging many other commercial products. Think of mechanized canning—culminating in the double-seamed cylinder of the "sanitary" can-making machinery of 1904—and mechanized bottle and cap making from the late 1890s. New forms of sugar consumption appeared with the invention of soda fountain drinks. Coca-Cola was first served in drug stores in 1886 and in bottles by the end of the century, and in the 1890s the mixing of sugar with bitter chocolate led to candy bars, such as Hershey's in 1900. Packaged pleasures of this sort—offered in conveniently portable portions with carefully calibrated constituents—allowed manufacturers to claim to have surpassed the sensuous joys of paradise. Chemists also began to be hired to see what new kinds of foods and drugs could be synthesized to surpass the taste, smell, and look of anything nature had created. A new discipline of "marketing" came of age about this time—the word was coined in 1884—with the task of creating demand for this riot of new products, decked out increasingly in colorful and striking labels with eye- and ear-catching slogans.

New technologies also sped up our consumption of visual, auditory, and motion sensoria. In 1839 the Daguerreotype revolutionized the familiar curiosity of the camera obscura—a dark room featuring a pinhole that would project an image of the outside world onto an interior wall—by chemically capturing that image on a metal plate in a miniaturized "camera" (meaning literally "room"). While these early photographs required long periods of exposure to fix an image, that time dramatically declined over the course of the century, allowing by 1888 the amateur snapshot camera and only three years later the motion picture camera. The effect, as we shall see, was a sea change in how we view and recollect the world. Sound was also captured (and preserved and sold) about this same time. The phonograph, invented in 1877 by Thomas Edison, became a new way of experiencing sound when improved and domesticated. And Emile Berliner's "record" of 1887 made possible the mass production of sound on stamped-out discs, capturing a concert or a speech in a two- or three-minute record available to anyone, anywhere, with the appropriate gear.

Access and speed took another sensual twist when a Midwesterner by the name of La Marcus Thompson introduced the first mechanized roller coaster, in 1884. Bodily sensations that might have signaled danger or even death on a real train were packed into a two- or three-minute adventure trip on a rail "gravity ride." Adding another dimension to the thrill was Thompson's scenic railroad (in 1886) with its artificial tunnels and painted images of exotic natural or fantasy scenes. This was a new form of concentrated pleasure, distilling sights and sounds that formerly would have required days of "regular travel." Rides, in combination with an array of novel multisensory spectacles, were concentrated into dedicated "amusement parks," offering a kind of packaged recreational experience, accessible (very often) via the new trolley cars of the 1890s. Some of the earliest and most famous were those built at Coney Island on the southernmost tip of Brooklyn, New York.

Innovations of this sort led us into new worlds of sensory access, speed, and intensity. Distance and season were no longer restraints, as canned and bottled goods moved by rail, ship, and eventually truck across vast stretches of space and climate—with mixed outcomes for human health and well-being.

Some of these new technologies nourished and improved our bodies with cheaper, more hygienic, and varied food and drink; others offered more convenient and effective medicines and toiletries. Still others provided unprecedented opportunities to enjoy the beauty of nature (or at least its image), along with music and new kinds of "visual arts." Amusement rides gave us (relatively) harm-free ways of experiencing the ecstatic and the exhilaration of danger—plus a kind of simulated or virtual travel; photography froze the evanescent sight, preserving images on a scale never previously possible, and with near-perfect fidelity. Yet packaged pleasures also led to new health and moral threats.

In the most extreme form, concentrating intoxicants led to addictions—physical dependencies that often required ever-increasing dosages to maintain a constant effect, and substantial physical discomfort accompanying withdrawal. Here of course the syringe injection of distilled opiates is the paradigmatic example, and addiction to tobacco and alcoholic drinks must also be included. But the impact of concentrated high-energy foods is not entirely different. Fat- or sugar-rich foods produce not just energy but very often endorphins, morphine-like painkillers that offer comfort and calm. That is one reason they are called "comfort" foods. These rich foods cause neurotransmitters in the brain to go out of balance, resulting in cravings.[17] By contrast, the natural physical pleasures of exercise are much less addicting because we get tired; and some "excess"—here pain is gain—can actually make us healthier.

Not all packaged pleasure dependencies were so obviously chemical. Engineered pleasures often create astonishment and delight when first introduced, for example, but can also raise expectations and dull sensibilities for "unpackaged" stimuli, be they nature's wonders or unaided convivial and social delights. The pleasures of recorded sound, the captured image, and even the amusement park ride and electronic game often satisfy with a kind of ratcheting effect, rendering the visual, auditory, and motion pleasures in uncommodified nature and society boring. In this sense, the packaging of pleasure can turn the once rare into an everyday, even numbing, occurrence. The world beyond the package becomes less thrilling, less desirable. In the wake of the telephoto lens and artful editing of film—with all the "boring bits" taken out—nature itself can appear dull or impoverished. Why go to the waterfall or forest if you can experience these in compressed form at your local zoo or theme park? Or on IMAX or your widescreen, high-def TV? Packaged pleasures of this sort may not induce physical dependencies, but they can create inflated expectations or even degrade other, less distilled or concentrated, kinds of experiences.

Another point we shall be making is that packaged pleasures have often *de-socialized* pleasure taking. Many create neurological responses similar to those of religious ecstasies, physical exercise, and social or even sexual intercourse, and can end up substituting for, or displacing, such enjoyments. Weak wine and mild natural hallucinogens have long enhanced spiritual and social experience, but the modern packaged pleasure often has the effect of privatizing satisfaction, isolating it from the crowd. Think of the privatization of public space through portable mp3 players, or the isolating effect of television.

The key point to appreciate is that we today live in a vastly different world from that of peoples living prior to the packaged pleasure revolution, when a broad range of sensual pleasures came to be bottled, canned, condensed, distilled, and otherwise intensified. The impact of this revolution has not been uniform, and we acknowledge and stress these differences, but it does seem to have transformed our sensory universe in ways we are only beginning to understand.

The packaged pleasures we shall be considering in this book include cigarettes, candy and soda pop, phonograph records, photographs, movies, amusements park spectacles, and a few other odds and ends.

But of course not all commodities that are tubed, packed, portable, or preserved can be considered packaged pleasures. For our purposes, we can identify several key and interrelated elements:

1. The packaged pleasure is an engineered commodity that contains, concentrates, preserves, and very often intensifies some form of sensual satisfaction.
2. It is generally speaking inexpensive, easy to access (readily at hand), and very often portable and storable, often in a domestic setting.
3. It is typically wrapped and labeled and thus often marketed by branding. Although often portable, in the case of the amusement park, it can also be enclosed and branded in a contained and fixed space.
4. The packaged pleasure is often produced by companies with broad regional if not national or even global reach, creating a recognizable bond between the individual consumer and the corporate producer.

Of course we are well aware that many other consumer products exhibit one or more of these attributes—clothes, cars, books, packaged cereals, cocaine, pornography, and department stores just to name a few. Our focus will be on those packaged pleasures that signal key features of the early part of this transformation, and notably those that involve the elements of containment, compression, intensification, mobilization, and commodification. And we recognize that we will not offer an encyclopedic survey of pleasures that have been intensified and packaged—we won't be treating the history of pornography or perfume, for example, and will consider narcotics and alcoholic beverages only briefly.

We should also be clear that the packaged pleasure revolution is ongoing and in many ways has strengthened over time, as pleasure engineers find ever-more sophisticated ways of intensifying desire. And we'll consider this history at least briefly. Since funneled fun has a tendency to bore us over time, pleasure engineers have repeatedly raised the bar on sensory intensity. Nuts and nougat were added to the simple chocolate bar, and cigarette makers added flavorants and chemicals to enhance or optimize nicotine delivery. The visual panel in motion pictures has been made more alluring with increasingly rapid cuts, and recorded sound has seen a dramatic expansion in both fidelity and acoustical range. Roller coasters went ever higher and faster while also becoming ever safer. Pornography is delivered with ever-greater convenience and is now basically free to anyone with an Internet connection. Even opera fans can now hear (and see) their favorite arias with a simple click on YouTube—at no cost and without leaving home (or sitting through those "boring bits"). Entertainment without the "fiber," one could say.

Another outcome of the packaged pleasure revolution, then, is the progressive refinement—really reengineering—of sensory experience in the century or so since its beginnings. Optimization of satisfactions has become a big part in this, as one might expect from the fact that packaged pleasures are very often commodities produced by corporations with research and marketing departments. Menthol was added to cigarettes in the 1930s, with the idea of turning tobacco back into a kind of medicine. Ammonia and levulinic acid and candied flavors of various sorts were later added to augment the nicotine "kick," but also to appeal to younger tastes. Flavor chemists meanwhile learned to manipulate the jolt of "soft drinks" by refining dosings of caffeine and sugar, while candy makers developed nuanced "flavor profiles"—surpassing traditional hard candy, for example, with the sensory complex of a Snickers.

Optimization and calibration we also find in other parts of this revolution. The intense thrill of a loop-de-loop ride, debuted first at Coney Island in the 1890s, gave way to the more varied sensuality of "themed" rides. Roller coasters have been designed to go to the edge of exhilaration, stopping just short of the point of nausea or injury. The same principle works with gambling, where even losers keep playing because of the carefully calibrated conditioning that comes with the periodic (and precisely calculated) win built into the game. Pleasure engineers have learned how to create video games that are easy enough to engage newcomers, but complex enough to sustain the interest of experienced players. Gaming engineers even seek to encourage (or require) physical movement and social interactions—think Wii games—to counter critics cautioning against the bodily and social negatives of overly virtualized lives.[18]

Our focus is on the origins of the technologies involved in such transformations, though we also are aware that such novelties have always encountered critics, those who worry that an oversated consuming public would lose control and abandon work and family responsibilities. But the reality in terms of social impact often has been quite different. Few of these *optimized* pleasures have ever undermined the willingness of consumers to work and obey—and have done little to undermine nerves and sensibilities (as some have feared). Indeed they have often contributed to a new work ethic driven by new needs and imperatives to earn and toil evermore in order to be able to afford the delights of movies, candy, soda, cigarettes, and the rest of the show. Over time, and often a surprisingly short time, these commodified delights have become a kind of second sensory nature—customary and accepted ways of eating, inhaling, seeing and hearing, and feeling.[19]

Scholars have long debated the impact of "modern consumer culture," albeit too often in negative terms without considering the historical origins of the phenomena in question. In the 1890s, the French sociologist Émile Durkheim feared that the "masses" would be enervated, even immobilized, by technical modernity's overwhelming assault on the senses. And Aldous Huxley in his *Brave New World* (1932) warned of a coming culture of commoditized hedonism oblivious to tyranny.[20] Jeremiahs of this sort have singled out different culprits, with blame most often placed on the "weaknesses" of the masses or the manipulation of merchandisers, with the hope expressed that the virtuous few in their celebration of nature and simplicity would constitute a bulwark against immediate gratification and degrading consumerism.[21] These critics have been opposed by apologists for "democratic access" to the choice and comforts of modern consumer society—who champion the idea that only killjoy elitists could find fault in the delights of pleasure engineering. This perspective dominates a broad swath of social science—especially from neoclassical economists (think of George Stigler and Gary Becker's famous dictum on the nondisputability of taste).[22]

We argue instead that we need to abandon the overgeneralization common to both jeremiahs and free-market populists. Of course it is true that the very notion of a "packaged pleasure revolution" suggests certain links between the cigarette, bottled soda, phonograph records, cameras, movies, and even amusement parks. But the impact of these various inventions over the decades has been very different, and cannot be subsumed under some procrustean notion of "modern consumer culture." Rather, as we shall see, their distinct histories suggest very different effects on our bodies and our cultures that would seem to require very different personal and policy responses. Our view is that the sale of cigarettes (as presently designed) should be heavily regulated and ultimately banned, for example, while soda should probably only be shamed and (heavily) taxed. And we make no policy recommendations for film or sound "packages." But we certainly need to better understand how these technologies have shaped and refined (distorted?) our sensibilities.

We should also keep in mind that there are global consequences to the packaged pleasure revolution—and that most of these lie in the future. This is unfinished business. Overconsumption is part of the problem, as is the undermining of world health (notably from processed sugar and cigarettes). The revolution is ongoing, as the engineered world of compressed sensibility spreads to ever-different parts of the globe, and ever-different parts of human anatomy and sociability. It may be hard to opt out of or to escape from this brave new world, but the conditions under which it arose are certainly worth understanding and confronting.

This book takes on a lot. Our hope is to move us beyond the classic debate between the jeremiahs against consumerism and the defenders of a democratic access to commercial delights. We root mass consumption in a sensory revolution facilitated by techniques that upset the ancient balance between desire and scarcity. We take a fresh look at how technology has transformed our nature.

# Containing Civilization, Preserving the Ephemeral, Going Tubular

Nature is ephemeral—at least that part that grows and dies. When plucked, a plant will spoil or simply disappear. The sweet red apple turns mushy, just as warm meat becomes rancid. Unpreserved on earth, on a human scale at least, light, sound, and smell all come and go in a flash. Before investigating our central topic, the packaged pleasure revolution, we must consider this broader question: how did humans first learn to contain nature's many gifts of materiality and sensuality?

One key stage in the rise of human civilization was the discovery of ways to preserve and store fruits of the harvest. Hunter-gatherers had typically had to enjoy (and replenish) themselves whenever opportunity arose—and starvation must have been a perennial threat. Preserving food changed that. Among the other markers of early humanity (tools, speech, art, and religious rites) was this push to plan for times of scarcity, a capacity eventually made possible by the invention of baskets, bags, pottery, bowls, and other devices for containing and mobilizing nature. Containerization liberated us from nature, at least a little. This is most obvious with food. Neolithic peoples beginning ten millennia or so ago learned to pack and preserve their nourishment, saving it from decay and also creating thereby entirely new kinds of foods—and sensory delights—in the process. Fermented drink is one notable out-come. Containerization allowed foods (and drink) to become portable while also being saved for use another day.

Pots, bags, and baskets—and eventually bottles—were central to early preservation technologies. Innovation in this realm was a long and drawn-out process extending across millennia, and one that does not easily give up all of its clues. Early containers were no doubt made from perishable materials such as leaves or grass or leather, leaving little or no trace in the archaeological record. Later technologies of ceramics and glass radically expanded human capacity to preserve nature and made possible the great ancient civilizations of China, India, Sumer, and Mesoamerica. Changes along these lines accelerated after the Industrial Revolution, especially in the generation around 1900. Indus-trial containerization made it possible to distribute foods throughout the globe; think only of what it would be like to live in a world without tin cans, cardboard cartons, and bottled drinks. Or bullet shells, lip-stick, toothpaste, cigarettes, and piping of various sorts.

As disparate as these various forms of "tubing" might seem, what unites them are certain shared engineering and conceptual origins—and for the subclass we shall focus on, certain similar effects on pre-serving, and often intensifying and delivering personal control over, an otherwise fleeting and diffused nature. Tubing helped transform cer-tain parts of nature into commodities, products that could be packaged and thus labeled, identifying and eventually advertising their contents. Suitably wrapped, tubing transformed the product by the color, image, and text of its external appearance, making the product something more than and different from its interior nature. And since the tubu-larized product could often be mass produced, what usually began as goods reserved for a privileged few came to be accessible to the broad population.

## Containing Civilization

Humans have long associated food with community, perhaps because finding and preserving it required cooperation. Hunters (and gather-ers) of meat must have shared with gatherers (and hunters) of plants.

No one knows how prestige may have been allocated to success in such endeavors, but we do know that most humans relied mostly on the berries, seeds, and vegetables provided by plant finders. And this practice of sharing survived the coming of agriculture as evidenced in our word "companion"—the breaking of bread together. Even in modern America, foodways are passed from generation to generation, often surviving the loss of an ancestral language (a love of pasta continued in second generation immigrants from Sicily even when knowledge of the Italian language did not). For thousands of years, humans have communed with the dead (or gods) by eating and drinking with them. The pleasures of eating are deeply social.[1]

Preserving food also required cooperation. Humans fifty thousand years ago probably used naturally occurring materials—including animal skins, horns, bamboo, large gourds, and leaves—to carry foods over distances. Prehuman hominids may have done the same for hundreds of thousands of years before that. Perhaps as long as twenty thousand or thirty thousand years ago, containers were made from modified natural materials—baskets woven from reeds, for example, or leather bottles or bladders with ties. Carrying food in containers or suspended from poles made possible the communal meal, allowing also certain economies of scale. So whereas uncooperative chimps might spend 70 percent of their time finding food, human hunter-gatherers might have to look for food only a few hours per day, opening up time for other cultural pursuits. The shift to a diet of meat also brought sustenance in a more concentrated form than fruits, vegetables, and grains. Domesticated animals could also be kept for sustained use, yielding milk or blood through a kind of ready-made "bottle," and with communal sharing even saved the need to gorge. Sharing also reinforced community, insofar as group participation made it easier to manage livestock and to dry, salt, or smoke meats to forestall spoilage.

Spoilage must have been one of the world's first great mysteries, prompting efforts to discover how to stop unwanted rot. Many different techniques have been invented, which must be considered among the grandest of all human inventions. Many millennia prior to the discovery of infectious microbes effective techniques of staving off putrefaction had been discovered, probably by trial and error. Neolithic peoples learned how to preserve meats and fruits in honey, maltose (from germinated barley), manna (from an insect that lives off the tamarisk bush), salt, and even maple syrup. Techniques such as these—and no doubt countless others—retarded the bacterial growth and molds that cause foods to spoil. Vinegars derived from fermenting sugars in grapes, fruits, and grains also impeded bacterial growth, as did a cold or a dry environment or salting. Methods varied by climate: strips of meat were hung out in the desert sun, Norwegians air-dried fish, and Iron Age Scots buried butter, meat, and even seed corn in the ground. Curdling milk into cheese was common in the west, just as fermenting fish (with salt) was common in the east. Many of our modern foods and flavorings are creations of such efforts to prolong "shelf life": think of yogurt, cheese, fish sauces, pickles, bacon, ham, many kinds of bread, and all alcoholic beverages. Food itself was sometimes used as a container, as when Europeans in the Middle Ages used bread as jars and butter as a sealant to preserve meat and vegetables. Cornish pasties, with their contents of cured meats, onions, sliced turnips, carrots, or even cooked jam preserved in a thick crust of bread served as a midday meal for English field hands. And the pastry crusts of meat and fruit pies served a similar purpose.[2]

Much earlier, however, the most efficient form of containing and preserving food and drink was the ceramic pot. Clay pots and figures date from at least twenty-five thousand or even thirty thousand years ago, but the invention of the potters' wheel in the Near East between 6000 and 4000 BCE significantly increased output and quality of the clay-based container, a key accompaniment of emerging urban life. Glass, invented about 3500 BCE in Mesopotamia and widely produced in Egypt about 1500 BCE, was made by melting and molding easy-to-find materials—including potash or soda (from ashes of wood or seaweed), silica (usually from sand), and eventually limestone for hardening. Early glass cups, bowls, and jars were fashioned by coiling ropes of molten silica and from 1200 BCE by pressing the liquid material into molds. Glass blowing was invented by Syrian craftsmen sometime between 27 BCE and 14 CE, allowing skilled artisans to make a hollow tubular vessel that was also impermeable: the glass bottle. Glassmaking was a lost art in Europe after the Roman Empire but was preserved in

THEBAN GLASS-MAKERS.

**Figure 2.1** Ancient Egyptian glassblowers from *Harper's New Monthly Magazine*, February, 1871, 338.

the Islamic world and then revived in Europe during the Italian Renaissance.[3]

Why were early containers so important? The first key fact is that artificial containers made it easier for people to create quasi-permanent settlements, leavening also the perpetual search for nourishment. Interesting also, however, is that pottery seems to have predated agriculture and may not originally even have been utilitarian. There is some evidence to suggest that early clay pots were used to represent or contain the remains of ancestors. Pots can also be seen as a concomitant of the move by humans to "pot" themselves in artificial housing. Pots were part of the process of human self-containment in shelters, much of which must have taken place prior to routine animal and plant domestication. In turn, agriculture and the accumulation that it hastened certainly encouraged containerization and sedentary life.[4]

Containerizing food was also important because it helped to create privileged classes and eventually to undermine the communal nature of food gathering and preserving. As Rousseau in the eighteenth century well understood, the ability to preserve a surplus led to private property and to privileges for those controlling the most productive land and largest surplus. This power did not always come from conquest or theft, despite claims of anarchists such as Pierre-Joseph Proudhon. More likely is that elites acquired power from their roles as overseers of the distribution of communal stores, especially of food. Hunters of fish, reindeer, grass seeds, and much else worked more efficiently when organized into groups to dry and preserve these goods. Aggrandizers must have competed for control over this surplus and the collectivities that made it possible. Power and prestige could be consolidated by controlling food stores used to fuel public festivals or emergencies.

Gradually, however, as scarcities and insecurities became less severe, elites must also have begun using surpluses for their own personal needs and to display prestige to an exclusive group. This is another point where containers came into play. Along with art, beads, and other ornamentation, elites often used pots and jars to advertise their authority and dispense favor. Feasts were used to impress potential allies; this could involve elaborate presentation of decorated pottery and jars, stimulating further demand for ceramics and glass. Honor came in part from the fact that display containers of this sort required a great deal of labor — often by slaves trained in specialized crafts. And when ceramic and glass bowls and jars became more widely available, the elite moved on to new expressions of prestige (gold, gemstones, and furs, for example).[5]

Elite dominance advanced further when aristocrats learned to control exotic goods by their monopoly of external trade. Obtaining imported wine, for example, required social networks that nonelites might not have. Emerging aristocracies had the means to engage in competitive feasting, gift giving, and accumulation of treasured goods. This rivalry must have led to challenges for control over these resources in war and feuds, but it also contributed to the development of more sophisticated containers.[6]

Communal feasts survived, of course, but prestige display made even private pleasures more desirable. Only with the democratization of cans and bottles in the second half of the nineteenth century did the ceramic container lose its near exclusive association with the rich (vs. the pile and pit of the poor, or common crock or keg).[7]

Another impact of early containerization is perhaps less obvious, but especially germane to our topic. Pots and bottles created the new and "unnatural" sensual experiences of cooked and mixed food. Claude Lévi-Strauss long ago proposed cooking as one of the fundamental acts dividing nature from culture (he in fact thought it the most basic). The human ability to use fire helped to create what Michael Pollan has called the omnivore's dilemma—what amongst so many possibilities should we eat?—by vastly increasing the range of edible plants and animals. Cooking killed poisons and made food easier to digest, an advantage not needed by ungulates like cows and sheep with their long and complex guts.[8] Roasting food over an open fire was the most obvious form of this cultural act but containers kicked this up a notch, making possible for the very first time boiling, frying, and baking, all of which must have been discovered soon after the invention of ceramic containers. Japan is home to the oldest known ceramic cooking vessels, with the earliest so-called Jōmon vessels dating to about 13,000 BCE. Similar techniques passed from China to the Near East and Europe between twelve thousand and six thousand years ago. Some anthropologists suspect that cooking vessels, along with boiled and baked foods, were mainly invented by women, but no one can be sure.

One thing we do know is that cooking in ceramic vessels made many formerly inedible seeds, vegetables, fruit, shellfish, and snails palatable, even delectable. We also know that cooking meats in pots (vs. roasting over an open fire) saved juices that could be used for broths or for flavoring vegetables and cereals—whence the origin of soups and porridges. Pot cooking would eventually spread across the globe, though we also know of periods of atavism, as when medieval European elites reverted to gorging on roasted game, symbols of their power and prowess as warriors and hunters,[9] bits of which survive in the outdoor barbecue, one could argue.

We also know that hunting and herding were often denigrated by agricultural empires as backward or primitive. Ancient Greek and Roman writers, living in civilizations relying heavily on agriculture and ceramics, frequently derided "barbarians" dependent on roasted meat and dairy. The *Iliad* and the *Odyssey* honor civilized men as "bread eaters." *Gilgamesh*, the Mesopotamian epic from four thousand years ago, tells similarly of a "wild man" who subsisted on raw food but was transformed into a "human" when a woman gave him bread and its fermented compatriot, beer, both of which required contained heating.

## Tubing Euphoria

An inevitable extension of the pot-cooked meal is the transformation of perishable nature—often in the form of grain—into alcohol with the fermentation of plant sugars. This was easily the cleverest exploitation of the fact of plant decay following harvest. Beer today is made from partially germinated barley, that is, malt, cooked in water to which yeast is then added to cause fermentation. But the very first beer must have come from the happy accident of tasting grain too long untended. Hops were added much later—in the Middle Ages—to give stability and a tangy taste. Many other seeds and plants have been used to make "beer," including wheat, oats, pumpkins, and even artichokes and green cornstalks (as was common in colonial North America). Geography and climate shaped the choosing of a particular poison: wine, for example, was part of the trade economy of the ancient world because (unlike beer) it could be stored and transported relatively easily from specialized centers of cultivation. Appearing in the Zagros Mountains possibly as early as 5400 BCE and in Georgia apparently even earlier (around 7000 BCE), grape wine had spread to Egypt by 3000 BCE and became vital in the economies of classical Greek empires. Wine arrived on the southern coast of France in the fifth century BCE, on boats laden with amphorae.[10]

The discovery of wine cannot have been too difficult, as wine is generated from the yeast found naturally on the skins of grapes. This yeast transforms grape sugar into alcohol, which means that whoever first stored grapes (or grape juice) in jars or pots for any length of time must have noticed the transformation. Vintners subsequently sped the process by using starter cultures of yeast to encourage the growth of favorable microbes.[11]

We today tend to think of alcohol as a recreational intoxicant: the ritual or "spiritual" aspects of alcohol are preserved even in etymology, since our word "alcohol" stems from an Arabic term meaning "the ghoul" or, literally, "the spirit." For most of human history, however, wine, beer, and the like were also considered basic or even indispensable foods. The anthropologist Michael Dietler estimates that between 15 and 30 percent of the grain produced over the history of human agriculture has gone into fermented drink. The English language alone boasts several hundred ways to express being drunk, and the container revolution made this linguistic efflorescence possible.[12]

Building on the techniques learned from fermenting grains and fruit, early chemical (and alchemical) engineers eventually also learned how to strengthen sensory impacts through distillation. Transforming herbs and flowers into bottled perfumes is a striking example of this push to preserve some of nature's most delightful sensations, including the sweet-smelling scents of the blossom. The very word "perfume" (from the Latin *parfumare*, "through smoke") betrays its origins as incense used in ceremonies, especially religious rituals. The key was to extract essential oils, liquids that would evaporate and effloresce in the form of a fragrant aerosol or "essence." Early methods included *enfleurage*, dating from ancient Egypt, a process by which fats such as lard and tallow would be pressed onto flower petals to absorb fragrant oils. The resulting *pomade* was dissolved in an alcohol-based solvent to obtain the essential oil. Another method was *expression*, involving the capture of precious oils from the skins of citrus plants in a press.[13]

The most common means of extracting essential oils, however, then as now, was *distillation*, and the basics were clearly known in ancient Greece and Rome. Distillation was further developed by medieval Arab alchemists, entering medieval Europe via Salerno (in southwestern Italy) in the eleventh century CE. A typical distillery by this time consisted of a sealable tank or "still" into which plant parts were placed along with water, onto which was then bolted a "head" topped by a gooseneck leading into piping of some sort, in which the resulting vapors would condense (by cooling) into liquid form. When the still was heated, the liquid would evaporate according to the volatility of the constituent vapors, much as we today divide petroleum into different "fractions" in an industrial cracking column. And since essential oils and water condense at different rates, oils of various sorts could be obtained in relatively pure "distilled" form. These purified oils could then be bottled and, if need be, sold.

By fortuitous accident, distilling could also be used to render wine into brandy. Here the crucial fact is that alcohol evaporates (and condenses) at a lower temperature than water, allowing the more potent (i.e., volatile) spirits (*esse*) to be extracted. Wine by virtue of its basic biochemistry cannot be fermented past about 15 percent alcohol, because yeast microbes cannot tolerate anything higher than this, which is why distillation was such a remarkable novelty. Distillation broke the bounds of naturally fermenting grains or grapes by allowing the production of drinks containing arbitrarily high levels of alcohol (through repeated distillation). And since distilled wine (as was discovered) could be set on fire, it was called "burnt wine" or, via the German, *brandy*.

The art and craft of distillation was so seemingly magical and supernatural that fourteenth-century Europeans treated the concentrated fragrance of perfumes and alcohol-rich spirits as revered medicines (sugar and spices were treated similarly, and for similar reasons). Brandy was also called *aqua vitae* ("water of life") and was sold in apothecaries' shops to cure everything from melancholy to the plague. Even in Victorian America, whiskey was kept in the homes of temperance families as a ready cure for "fainting spells."[14]

Distillation achieved a wider impact once Arab traders learned to grow sugarcane in the Atlantic Islands (for more on which, see chapter 4). By the mid-seventeenth century, tropical planters had discovered that molasses, a byproduct of refining sugar, could be distilled into "rumbullion" (English slang for a brawl). Shortened to "rum," this new and cheap liquor (typically around 48 percent alcohol) displaced brandy as the quick route to inebriation in both England and the colonies. Gin, a liquor distilled from fermented grain and flavored with juniper berries, appeared in Holland early in the seventeenth century. Gin was served to English soldiers in the Thirty Years' War and returned with them to the British Isles. Other distilled drinks were made from fruit: calvados from apples, for example, or kirsch from the pits and flesh of cherries. After the Revolution, Americans specialized in whiskey, made on the cheap from fermented cereal. Here again, innovation took the form of a kind of salvage operation. In the Ohio Valley in the 1820s, corn (i.e., maize) was grown in such superabundance that it had to be converted to "concentrated corn" as either whiskey or pork. Even in 1790, before this explosion of corn farming, the average

American drank more than twice what their compatriots do today (six gallons of alcohol a year for every American over fifteen years of age). And much of the country's early apple crop, especially from the seedlings provided by Johnny "Appleseed" Chapman, was pressed and fermented into hard cider (contra Disney). Some of this was then distilled or, to the same effect, frozen, separating water in the form of ice from the liquid alcohol, yielding a high-proof hooch called applejack.[15]

The creative energy going into such inventions was remarkable, and in Europe by around the year 1800 peasants, who had long made liquor from the residues of wine presses, cereals, and even fruits, had discovered that a potent brew could be distilled even more cheaply from potatoes. Potato schnapps, as a result, largely displaced home-brewed beer. Only at the end of the nineteenth century did large-scale breweries win back a more urbanized population by replacing traditional dark ale beer with lighter "lagers" (requiring cooler temperatures in fermentation).[16]

Brandy, rum, and whiskey all kept well or even improved with age. Distilled spirits were also cheaper and easier to ship than wine and beer, which is one reason so much international trade was done in the stuff.[17] Rum and brandy also did duty in the increasingly regimented armies of the seventeenth and eighteenth centuries. High-proof alcohol vastly increased ease of intoxication, a coping mechanism linked alternately to the speeding up of life and the perils (and chronic boredom) of military duties. Parallel in certain respects was absinthe, a liquor distilled from wormwood, a herbaceous plant (*Artemesia absinthium*) brought to France by soldiers fighting colonial wars in Africa. By 1850 absinthe had become the drink of choice among Parisian bohemians but also the poor because of its rock-bottom price. Claims that the "green fairy" (*la fée verte*) caused epilepsy, tuberculosis, insanity, crime, and even spontaneous human combustion led many countries to ban the drink. The United States outlawed absinthe in 1912, for example, eight years before Prohibition.[18] Adulterants caused most of the ill effects.

An upsurge of mind-altering consumption can be detected in the "long eighteenth century" stretching from 1660 to 1820, when Europeans (and American colonists) gathered up and reproduced these mind-altering substances from around the globe. A striking example is gin, consumed in near-epidemic proportions in England during the eighteenth century. In 1684, Englishmen and women consumed only half a million liters of gin; by 1737 this had risen to five million liters. Similar increases are recorded in other psychotropics — sugar-based rum of course, but also caffeine, sugar, and nicotine from the Middle East, Asia, and the Americas, with varying addictive effects. Psychotropics sometimes displaced more social activities, including festivals, pageants, and church rituals designed by religious and secular authorities to shape the moods and behaviors of subservient populations. During the long eighteenth century there was a fundamental shift in the social organization of euphoria from a culture of ritualized and seasonal festivals, often lorded over by elites, to a self-administered ecstasy found in the bottle.[19]

Of course early humans had long been skilled at finding and refining psychoactive substances throughout nature. The curious power of the opium poppy was discovered multiple times, and perhaps independently, from India to central Europe, with the earliest recorded use in Europe being sometime around 1600 BCE. For nearly three thousand years the plant's nutritious seeds and oil have helped to stave off hunger, fatigue, and pain as well as boredom for boatmen, peasants, and laborers. But while distilling fermented grain and fruit came early, it was not until 1804 that a German apothecary, Friedrich Sertürner, isolated morphine as the potent alkaloid of opium. This was a genuine medical triumph, allowing doctors to calibrate how much opium to administer, ending those less measured times during which doctors had been forced to give massive and dangerous amounts to alleviate pain. Morphine, by contrast, was chemically pure and thus measurable; it was also soluble in water, providing fast-acting relief with no gastric side effects. It was also more addictive, even as doctors began using (and abusing) it to treat all kinds of conditions, real and imagined. Among many others, the German chancellor Otto von Bismarck used morphine to gain relief from insomnia and to cope with digestive ailments.[20]

Key to the intensification of psychotropics was their delivery via the hypodermic syringe. The idea of using a glass or metal tube funneling into a needle to create a manipulable suction dates back to a ninth-century Muslim surgeon (Al-Zahrawi, in his book, *Kitab At-Tasrif*) who used such a device to remove cataracts from his patients' eyes.[21] In 1760, European physicians began using a hollow needle to inject medicines into the body (first into boils and other growths above the skin). But the decisive invention came in 1853, when Charles Gabriel Pravaz of France created, and the Briton Alexander Wood popularized, a practical hypodermic syringe. With its hollow steel needle fine enough to pierce the skin without damage, the syringe could deliver chemicals directly into the bloodstream in measured doses. Other improvements (including a graduated scale on the barrel, a glass piston and plunger, and ever-finer needles) made the syringe ubiquitous in hospitals and medical offices. Much later, in 1956, a plastic syringe became available, and in 1974 disposable models appeared.[22]

Despite fears of puncturing veins and infection, the syringe was often a superior way to deliver painkillers when compared to ingesting them as pills or dissolved in liquids like wine. The advantages were twofold: 1) medicines injected directly into the bloodstream were fast acting; and 2) injected medicines avoided side effects on the digestive system. And it's no surprise that "needle use," with its negative taint, would come to be closely associated with morphine—given that this distilled form of opium would gain much of its addictive potency only with direct injection. In the 1880s, as needle delivery of morphine was beginning to be associated with addiction, physicians started to demand that they alone should have access to the syringe. The dramatic upsurge in chemically induced pain relief (and pleasure delivery) caused a certain level of moral concern or even panic, with one worry being that a syringe could be used clandestinely, undetected. Medical triumphs gradually broadened use of the syringe, however, especially (from the 1880s) when it came to be widely used to deliver antitoxins and vaccines. This multiple use made the hypodermic needle a tube of great moral ambivalence, a condition it retains even today.[23] The related phenomenon of tubing in the form of the mechanically rolled cigarette—a kind of lung syringe for delivering nicotine—we shall deal with in chapter 3.

## Tubing Food and Drink

Tubular technology was taken to a new level with the Industrial Revolution; indeed, it is hard today to imagine a world without cheap bottles, "tin" cans, metal gun cartridges, toothpaste, hair cream, lipstick, and ready-made cigarettes, all of which came into being in the nineteenth century. There may even have been something inevitable about going tubular. Tubes were an efficient way of containing and often preserving stuff—a culmination of the age-old container revolution. But tubes also facilitated the confined and discrete flow of liquids and loose materials (like toothpaste, soda, or shredded tobacco), allowing new economies of scale in manufacturing. The tube used in manufacturing often ended in a tube used for consumption.

In fact, the mass production of goods in tubular form was itself an outcome of the geometry of mechanization. From the 1770s on, steam engines spurred a revolution in rotary power (replacing water wheels), allowing the dramatic spread of turning lathes, drills, and boring mills. Machines such as these were used to churn out the rollers, pipes, and other tubes that, along with molding presses and cutting tools like lathes and milling machines, were essential for modern manufacturing. None of this was as sensational or as graphic as the camera or the phonograph, but the tubular revolution was nonetheless impactful and, interestingly, linked to the rise of continuous process manufacturing. Factories outfitted with rotary machine tools were often sites of continuous motion, with products shaped by rollers, trip hammers and cam-activated dies, and moved by conveyers and gravity slides powered by belts and gears. As early as the 1880s, machine tools organized in this new fashion enabled a nearly automatic manufacturing process— with human labor often reduced to the role of maintenance and repair (along with invention, marketing, and sales).

Continuous-process manufacturing took some time to perfect, however. As late as the 1880s, for example, no one had yet figured out how to make the seams and seals necessary to transform sheet metal into a cylindrical can, and this remained a speed bump in the tubing revolution. We should also appreciate, though, that other components of continuous process manufacturing were already well-known by the mid-nineteenth century. Henry Brown's *507 Mechanical Movements*, published in 1868, was a treasure trove of illustrated information about how water wheels and steam engines could be harnessed to drive machine tools. Over the next two generations, machinists and inventors combined these and similar "mechanical movements" into the complex contrivances that would crank out consumer goods. By 1900, this unimpeded movement of raw materials through an array of cutting, shaping, and sealing devices dramatically accelerated the manufacture of cigarettes, bottled soda, canned goods, and countless other containerized consumables.[24]

One leading edge of this revolution was the development of the metal can, a radically new type of container formed not by the blowing of glass or turning of clay but rather by folding flat-sheet metal into a cylindrical form. Military competition had already led to the mechanized tubing of firepower (as in the development of cannons, rifles, and the bullet cartridge, or "shell"), but military demands would also spur the canning of foods. The principal catalyst here was the series of wars waged from 1792 through 1815 between revolutionary France and most of the crowned heads of Europe, which ended with Napoleon's defeat. In 1795, in one of several attempts to find new ways to feed the hundreds of thousands of soldiers massed for long and distant campaigns, the French government offered F12,000 to any inventor who could come up with a cheap and effective means of preserving food for soldiers on the go. False starts were many, but in 1809 a Parisian confectioner by the name of Nicolas Appert won the prize with his sealed food jar. The "father of canning" had found that fruits, vegetables, eggs, meats, and even fish, when boiled and stuffed into wide-mouth glass bottles sealed with a cork, would not spoil for months or even years. This simple solution was the culmination of Appert's efforts to find an alternative to traditional preservation techniques—such as drying and salting—that so often left foods leathery, tough, or acidic.[25] High costs made Appert's invention of limited value for Napoleon's expeditions, but an Englishman by the name of Peter Durand replaced breakable glass bottles with his newly invented metal "cans" in 1810. Named after the traditional cane "canisters," Durand's iron cans were covered in a thin coat of tin to prevent rusting—whence the name "tin can." By

1813, he was selling canned meat to the Royal Navy for sick sailors on ship and as a backup for all on long oceanic voyages. Tin cans assisted in the provisioning of explorers and long-distance seafarers, making it easier for European powers to conquer the world.[26]

The world's first cans were lead-soldered by hand. Artisans cut the rectangular body and discoid ends from sheet metal, bent the body around a cylindrical mold, and then soldered the seams along the sides and one end. Cooked food was then poured in through the open end, following which the top would be soldered into place, leaving only a small hole to evacuate air during cooking. After cooling, the hole was finally soldered shut. Using these techniques, even a skilled artisan could produce scarcely sixty cans per day, and much solder was used.

Canning began in the United States in 1839, with early operations concentrated in New York and New England. Canned food from this era was expensive and earned a reputation for spoiling because manufacturers were often sloppy or ignorant of proper heating techniques. Heating to well above boiling was required—especially for nonacidic foods—which meant that canned food was often reduced to flavorless, unrecognizable chunks. The lead solder could also taint the contents or even poison the eater. To make matters worse, there were no can openers in these early years! Soldiers opened tins with their bayonets, and civilians used chisels. As late as the 1850s tinned food was at best a status symbol in Europe, with many considering it a frivolous novelty.[27]

In 1856, however, an American newspaperman by the name of Gail Borden advanced the art by evaporating water from milk and adding sugar, producing sweetened condensed milk in cans using a vacuum technique borrowed from the religious commune of the Shakers. Borden was so smitten by the idea of compressing food into packages that he declared, "I mean to put a potato into a pillbox and a pumpkin into a tablespoon."[28] Condensation (by evaporation) made possible the distribution of otherwise perishable milk far from its bovine and rural origins, and offered farmers a profitable alternative to converting milk into cheese for sale beyond their immediate environs. Canners also learned how certain salts could be added to the contents, making it easier to raise the temperature during cooking. Cans were widely used to provision soldiers in the Civil War, albeit mostly those on the

Union side, and mostly just with pork and beans and condensed milk. Borden also, though, used his machinery to condense blackberry juice for sick and injured soldiers. By 1862, California peaches, berries, and tomatoes were being canned and given to officers. Returning veterans with an acquired taste for tinned food introduced their families to this novelty, expanding the market. And canning got a further boost in 1874 when a Baltimore canner, A. K. Shriver, patented the retort pressure cooker, speeding the process and making it safer.[29]

Mechanical food processing expanded demand for what had once been "seasonal" foods, but it also increased (and regularized) supplies when acreage for crops and livestock grew with the "opening" of the American West. "Opening" of course is a euphemism given the violence of the process, which involved the forcible removal of the indigenous natives and the buffalo, a task largely finished by the end of the 1870s thanks to racism and gun cartridges. Plains Indians north and south were quickly replaced by cattlemen and their vast herds of steers (and wheat farms), all serving to slake the commercial thirst for meat. Key to the trade was also the famous connection of the west–east cattle trail ending at Abilene, Kansas, from which was extended a south–north railroad ending in Chicago's meatpacking centers in 1867, two years before the "golden spike" linked east and west in Utah. This new system of rails created a means for the efficient transfer of livestock and prepared meat to rapidly growing population centers of the north and east, forging crucial links in the modern American food chain.[30]

Capping (and crucially facilitating) these extraordinary events were innovations in refrigeration technology. Gustavus Swift in 1879, for example, instead of shipping livestock, delivered butchered meat— lighter by about two-thirds—in refrigerated railcars equipped with dry ice. This made the Union Stock Yards of Chicago and its animal disassembly plants a center for a substantial increase in American meat consumption.[31] From here we can also trace our disconnect between the grim reality of raising and butchering animals and our lust for burgers and cold cuts. Commercial meatpacking is one reason we are able to morally detach food from its origins in living nature, for better or for worse.

Faster travel and improved preservation combined to deliver many other goods from far-flung regions. The extension of railroads from Chicago to New Orleans gave Midwesterners southern fruits and vegetables in winter—while wheat, corn, and salted pork moved from the north down south. The nationwide fame of the Georgia peach was launched in the 1870s, thanks to the expansion of north–south rails. Fresh fruit from California was making its way east by rail as early as 1869, and by the end of the century even tender vegetables like lettuce, long a delicacy of the affluent, were becoming readily available, even in winter. Iceberg lettuce triumphed because it was easy to pack (first in ice, whence the name) and relatively impervious to the lengthy rail journey to East Coast towns from California. Fresh tomatoes could always be found growing somewhere in the United States, and improved transport meant that seasonality was less and less a factor in what could arrive on one's table to be eaten.[32]

Still, the problem of fruit and vegetables rotting in transit (or storage) had hardly vanished. Preservation remained a daunting issue, especially as distances between production and consumption increased. And it did not take much convincing to whet people's appetite for exotic and out-of-season foods. Who in February would choose a diet of beans, parsnips, and dried meat over tomatoes, cherries, and soft meat? Here again, canning came to the rescue.

Can making doesn't get the attention it deserves from historians, perhaps because many of the changes were small and incremental. (Dramatic mega-technologies like Bessemer furnaces and Watt steam engines are the more traditional overfocus.) Gadgets that stamped out the body and then tops and bottoms of cans appeared in 1847 and 1849, with patents duly awarded. Seam and lid soldering was automated in the 1850s with the invention of the Little Joker, a small tabletop machine that rotated can caps in a bath of molten solder for sealing tops and bottoms. And in 1849, the American Henry Evans developed combination dies for stamping and bending the edges of can caps. These and other innovations led eventually to a dramatic increase in speed and volume of manufacture, allowing annual U.S. production to leap from 5 million cans in 1860 to 30 million only a decade later. In the late 1870s, improvements in the Little Joker for "floating" solder on the ends of cans made it possible for two men with assistants to turn out 1,500 cans per day—causing also a decade-long (and losing) battle of can makers to save their jobs from mechanization. Innovations in the 1880s continued with a multiple can capper, mechanical labelers, automatic slitters for cutting sheet "tin" into strips, and an automated machine called the Locker, used to seal the side seam of the can. Seam sealing was an important early bottleneck, and in the early 1880s Chicago's Edwin Norton developed a machine that folded the seams and applied external soldering, increasing a single machine's capacity to 2,500 cans per hour. Improved versions from only a decade later could crank out 6,000 an hour.[33]

Successful canning of course required a convenient opener, another unsung invention of great practical consequence. First in 1866 came the key-wind "tear-strip," a method still used today on some sardine cans. Then in 1875 appeared the familiar "piercing wedge" opener.

The greatest change in canning, however, came with the invention of the so-called sanitary can. Perfected in 1905, the sanitary can eliminated most soldering by use of double seams assuring an airtight seal. The process was a marvel of industrial engineering, albeit (again) virtually unsung by historians of technology. Tin plate rectangles were cut and curled around a die and fitted tightly together, creating a cylinder with a double seam running lengthwise along the can (see figure 2.2). Rollers then flared out the ends of the cylinder, following which circular tops and bottoms would be stamped with a downward countersink. Tops and bottoms could thus be made to fit snugly onto the can body, affixed with a rubber compound (invented in 1896) in the curl of the seam. With crimped double seams replacing lapped side seams, only a thin coat of solder was applied on the outside of the can, safely away from the food contents. The "sanitary can" allowed a wide range of tinned food to reach urban populations, especially as rival processors introduced ever-cheaper and more attractive foodstuffs festooned with colorful labels and catchy brand names.[34]

A very different direction was taken by designers of flexible, compressible tubes of the sort now used for toothpaste. Soft metal tubes designed to extrude their contents were first used for artists' paints in 1841, but it took another half-century for a Connecticut dentist to apply this same principle to dental cream, marketed first and most famously

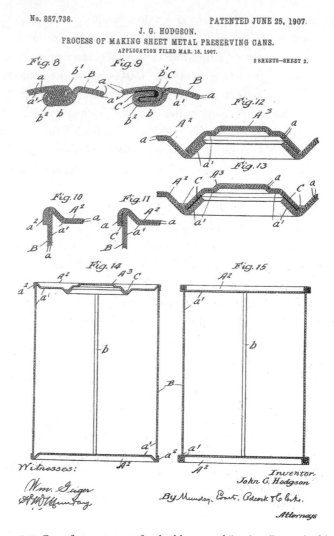

No. 857,736.                                PATENTED JUNE 25, 1907.

J. G. HODGSON.

PROCESS OF MAKING SHEET METAL PRESERVING CANS.

APPLICATION FILED MAR. 18, 1907.

2 SHEETS—SHEET 2.

**Figure 2.2** One of many patents for double-seamed "sanitary" cans, in this case the top seal. U.S. Patent Office, 1907 (no. 857,736).

by Colgate. Toothpaste tubes were made from metal until only a couple of decades ago in the United States, and similar tubes are still today used (especially in Europe) to dispense tomato paste and other soft foods, mayonnaise, and medicine.[35]

Of course even after the debut of the sanitary can, containers made from glass—the original "cans" of Appert—did not disappear, despite their disadvantage of breaking easily. Glass jars had other pluses: glass is chemically inert and won't corrupt its contents; moreover, thanks to millennia of glass blowing and ceramics, the making of bottles and jars was a well-developed craft. Indeed, little improvement occurred from the invention of the blowpipe in antiquity until the nineteenth century. Sealing the contents was rather simple with "corks" of various sorts; wood or even oil-soaked rags were in common use, and French wine makers in the seventeenth century started using plugs from the bark of the cork trees native to Portugal, Spain, and neighboring parts of Africa. But corking really only worked for narrow-mouthed bottles.

Overcoming the inconvenience and impracticality of cork for glass jars was an easy-to-miss invention by John Landis Mason from 1858: the wide-mouthed, screw-lid Mason jar. What simple invention has had a more lasting impact? The Mason jar continues even today to preserve everything from pickles and peanut butter to "put up" home preserves (especially jams and jellies, but also tomatoes and beans and so forth). The domestic making of such items is surprisingly recent: rural (and town) women did not commonly "can" such foods until early in the twentieth century, when an improved Mason jar became readily available for home use.[36]

From the point of view of technique, a trickier problem than jarring was how to seal bottles, especially bottles of bubbly soda pop, a craze that began in hotel and drugstore fountains in 1806 but spread into a broader consumer culture with factory-scale mechanized bottling (see also chapter 4). The problem here was not so much spoilage or even resealing (since most people consumed the whole bottle in one sitting), but rather how to preserve the fizz before opening. After all, bubbles were the main attraction. One early solution was developed by the German-born Swiss watchmaker Johann Schweppes in 1794. His "lay down" bottle with a rounded bottom kept the cork wet and tight, thus helping to keep the fizz in. Improved caps included Henry Putman's wire clamp cork stopper of 1857. Even more ingenious was the internal "gravitating stopper" of John Matthew, crafted in the 1860s. After a bottle was filled and then inverted, an internal stopper dropped into the bottle's neck, sealing in the contents. Charles G. Hutchinson added to this mix a spring-type internal stopper in 1879. Here the consumer pushed down on a flexible looped wire attached to a rubber disk stopper inside the bottle (at the top), releasing the stopper without its falling into the bottle. When pushed in, Hutchinson's stopper made a "pop" noise, which is where we get the name "soda pop." The earliest Coca-Cola bottles in the mid-1890s were of the Hutchinson type.[37]

These diverse solutions for the problem of pressurized containment took a dramatic and ultimately decisive turn in 1892 with William Painter's invention of the "crown bottle cap." Painter that year in Baltimore patented a cork-lined cap that was crimped onto the top of a bottle made to have a specially formed lip. Resembling the British Queen's crown in miniature, Painter's crown cap quickly became the standard for the nascent "soda pop" industry, out-competing some 1,500 alternative designs.[38] It is still in wide use even today.

Just as important, though, was Michael J. Owens's invention of the automatic bottle blowing machine (conceived in the 1890s, with a patent awarded in 1904). Owens's device blew gobs of molten glass into bottle shapes through a series of rotary-arranged "arms" that worked rather like bicycle pumps. Molten glass was first sucked into a mold to make the bottle's neck, following which a calibrated burst of air blew the glass to form the hollow body of the bottle. The effect was a massive increase in speed of manufacture: instead of the two hundred or so bottles that a traditional glass blower could issue in a twelve-hour day, Owens's machine produced up to fifty-seven thousand per day — and all virtually identical. And so when Benjamin Thomas and Joseph Whitehead won the rights to bottle and to distribute Coca-Cola in 1899, they utilized the Owens machine to make their (now famous, and famously distinctive) bottles. A standardized design was adopted in 1916, and by the 1920s more than a thousand authorized Coca-Cola bottling plants were in operation, producing over a billion bottles per year.[39]

Once objects of luxury, the glass bottle and metal can by the end of the 1800s had become ordinary, even throw-away, conveniences. And it would be hard to overestimate their impact on everyday life. These containers helped make it possible to dine virtually anytime on foods previously available only a few weeks a year, while simultaneously en-

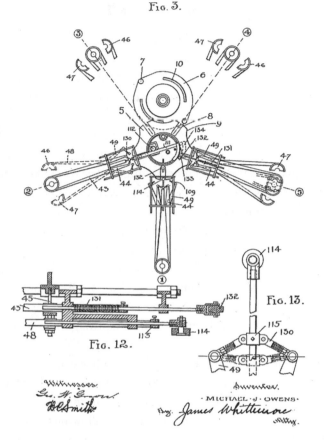

No. 766,768.

PATENTED AUG. 2, 1904.

M. J. OWENS.
GLASS SHAPING MACHINE.
APPLICATION FILED APR. 13, 1903.

NO MODEL.

10 SHEETS—SHEET 3.

FIG. 3.

FIG. 13.

FIG. 12.

Witnesses:

Inventor.

· MICHAEL · J · OWENS ·

By James Whittemore

Atty.

**Figure 2.3** A patent for glass-blowing machinery by Michael Owens. Each of the arms shapes a bottle. U.S. Patent Office, 1904 (no. 766,788).

larging the kinds of food and drink available. This liberation from nature's tyranny of the seasons was perhaps most evident when canning first appeared; today we tend to take such things for granted, forgetting their revolutionary impact. Cans and mechanically blown bottles vastly extended the virtues of ancient pots, enlarging their range and convenience of access, spawning new tastes and nutritional forms. Tubing

the natural world transformed sensation, time, and space, with consequences we are only just beginning to appreciate.

## Paper and Plastic Tubes

New ways of working glass and metal made possible the modern packaged pleasure, but paper and later plastic were arguably just as important. Of course, paper had long been a form of wrapping and preserving nature. Paper is basically a sheet of crushed cellulose derived from
plants, screened into a planar sheet from a floating mass of pulp. In
China as early as the second century BCE, sheets made from mulberry
bark fibers were used to wrap foods — with other (later) uses ranging
from toilet paper and paper money to firecracker wrappings and pornography. Cellulose fibers were often derived from flax, from which
linen was also spun and woven. In the Latin Middle Ages, Chinese
techniques of paper making were transported into the Mediterranean
basin, from where they were then introduced into Europe, reaching as
far north as England in 1310. Paper for many years was a relative luxury,
though a paper-making machine invented in 1798 in France lowered
costs enough for the stuff to be used as a packaging material, eventually giving rise to the cardboard carton, a lightweight replacement for
the wooden box.

Paper did not become the cheap and familiar consumer good we
know until the second half of the 1800s, however, thanks to a series
of little-explored technical developments. Not until 1867, for example,
could cellulose fiber be made from wood pulp, allowing trees to replace
cloth and flax as the primary source of pulp for paper. Newspapers and
books became more affordable, but lowered costs also led to a revolution in "flexible packaging." The paper bag was invented in 1844 in
England, and in 1852 the American Francis Wolle introduced the first
of many paper bag–making machines. By the late 1860s an ingenious
square-bottomed bag was available, increasing its carrying capacity and
stability. Paralleling the development of bottle making and sanitary can
manufacture, printed paper bags were produced in a continuous process line beginning in 1905.[40]

Paperboard and cardboard were equally important in the packaging revolution. The Chinese had invented a relatively strong and stiff form of paper in the seventeenth century, but the first cardboard carton appeared in England only in 1817. (Most of these early cartons were round or oval, given the difficulty in making clean and durable creases in rectilinear boxes.) Another common form of cardboard consisted of wavy or corrugated paper set between two stiff sheets of paperboard. Appearing first in the 1850s, corrugated cardboard became an excellent replacement for the heavy wooden boxes and barrels then being widely used to transport canned and bottled goods. Corrugated cardboard was strong, light, and cheap, and nearly a thousand patents for improving its use as a packaging material had been filed by 1900.[41]

Retail uses of paperboard were limited, however, until the development of breakfast cereals in the 1880s. Although invented in 1863 by James C. Jackson and promoted by vegetarian guru Sylvester Graham as an alternative to the common breakfast of meat and eggs, dry cereal took off only in the 1890s. This was when Battle Creek, Michigan, site of a Seventh Day Adventist sanitarium, became America's cereal central. Two young sanitarium employees, the brothers John Harvey and William Keith Kellogg, in 1877 introduced to their vegetarian patients a breakfast cereal made from ground zwieback and other hard breads. By 1894, John Harvey had come up with granose, a cooked wheat rolled into flakes, and two years later he introduced "corn flakes" to sanitarium patients. Breaking with his older brother, William Keith launched Kellogg's Toasted Corn Flakes in 1906, adding sugar and massive advertising. As important as the food itself was its novel container, a wax paper–lined paperboard carton designed to keep the fragile flakes from being crushed or going stale. Without moisture-proof packaging, toasted flakes of corn and wheat meal would have been impossible to sell. Other boxed foods had more the character of a sugary snack: the makers of Cracker Jack, a confection of popcorn, peanuts, and molasses, expanded production dramatically when they developed a waxed paperboard wrapping in 1898. Automatic boxing made for volume sales: the contents cost the company only 1¢ while the packaging cost 2¢ (and a box retailed for 5¢).[42]

Plastics also began to play a role in containing perishables, albeit mostly not until the twentieth century. Styrene was discovered in 1831, vinyl chloride in 1835, and celluloid in the 1860s, but none of these became really practical for packaging until the 1900s. Styrene, distilled from the balsam tree, was brittle and shattered easily. Only in 1933 was the process refined in Germany, becoming Styrofoam in the 1950s. Celluloid, derived from plant cellulose treated with nitric acid, was another important first-generation plastic, used early on as a cheap substitute for ivory billiard balls and in everything from false teeth, shirt collars, and shirt fronts to combs and corset stays. Only in 1914, though, did a celluloid derivative called viscose (first developed in France) turn into the famous wrapping material cellophane.[43] Cellophane tended to discolor and crack over time, however, and could burst into flames fairly easily (not surprisingly, given that nitric acid and cellulose were key components in smokeless gunpowder).[44] Cellophane was widely used to encase cigarette packs from the 1930s; the transparent wrapping was in fact so closely identified with cigarettes that a 1937 indictment of cigarettes was titled *Death in Cellophane*.[45]

Paper packaging solved a host of problems for manufacturers, reducing handling costs and waste (from vermin) while also saving retailers time, as they no longer had to measure from bulk containers (such as barrels). Paper packs and cartons also allowed goods to be distributed more easily, more cheaply, and over greater distances to people who usually had no idea who had made them or where they came from. Packaging also enabled the sale of smaller portions—pocket-sized packets of candy or cigarettes, for example—encouraging regular purchases, gifting, and impulse buying.[46]

## Labeling and Selling the Package

Nowadays, wherever there is a commercial package there is usually a label to identify its contents. But from its very beginning, the label was about much more than convenience. A package of pancake flour bearing the label "Aunt Jemima" meant something very different from a barrel of pancake flour refined at the local mill—even if the flour in the barrel might be chemically identical. The label signaled to its purchaser more than an identification of contents; an aura of authority or emotional attachment could also be conveyed, impacting buyer's choice and transforming how consumers related to retailers and retailers to jobbers and manufacturers.

Distinctive trade signs had long identified the businesses of butchers, bakers, and candlestick makers, and paper labels date from at least the fifteenth century. As early as the 1660s, local English merchants used these markers to separate their wares from imports they considered to be of inferior quality. And in the early years of the 1800s, retailers sometimes repackaged and labeled goods they had bought in bulk. But tinned, bottled, and boxed goods remained mostly hand crafted and expensive until the 1870s. Colorful labeling on a broad scale was not really even possible until the introduction of chromolithography in the 1840s, which meant that colored labels up to then were largely confined to luxuries featuring emblems of quality and pride. Attempting to reach a broader market, the Smith Brothers of Poughkeepsie, New York, pioneered a distinctive trademark to sell their cough drops in labeled glass jars in 1866. And in the 1870s the Library of Congress began registering retail trademarks, allowing manufacturers to identify their products with a distinctive and protected brand.[47]

Product labeling developed gradually, paralleling the emergence of packaging commerce itself. Labeling was part of the nascent discipline of marketing, invented principally to dispose of the massive surpluses heaped up by the packaged pleasures revolution. New machinery increased output, requiring manufacturers to increase sales in order to pay for greater costs of investment. One solution was to court consumers with labeling, which helped to create consumer demand. The historian Susan Strasser has shown how labeling also reduced the retailer's ability to sway the shopper's choice. Customers increasingly wanted the "reliable," nationally advertised "name brand," and retailers gradually lost their traditional power to favor house or local brands or generic commodities scooped from a barrel.[48]

Trademark law also emerged about this time, largely to protect against forgers, fakers, and infringers on branded goods. While trademarks were registered from 1870 on, the U.S. Supreme Court in 1879 rejected the claim of manufacturers that a brand name was private property. The United States Trade-Mark Association persevered, however, and by 1905 Congress passed a law recognizing trademark registration as evidence of ownership of a mark, making it illegal to counterfeit. Trademarks were also unlike patents and copyrights in that they never expired.[49]

Packaged goods almost always came with labels identifying their "special" contents. But what really made these labels work—meaning sell—was advertising, a kind of label-writ-large posted to the world. Ads that "hit" the eye with a powerful image and emotional appeal appeared in tandem with the name-brand product. Modern ads evolved from the trade card, a Victorian vehicle used to announce a product or retailer, something like the business cards handed out by professionals today. Though informational and plain at first, from the 1860s onward many were brightly colored and festooned with attractive pictures: iconic landscapes, patriotic displays, exotic visions of Native Americans or Japanese, images of famous opera singers or sports heroes or even images of children from around the world. Trade cards shared common themes with the commercial greeting cards pioneered by Louis Prang in 1873, but could also be placed in boxes of tea, coffee, soap, and cigarettes (to stiffen the paper packaging). Many retailers gave away cards to customers, encouraging their collection or even reuse as postcards. Trade cards appealed to children, and merchandisers enlisted the child's "pester power" to get a harried or indulgent parent to buy the item promoted with the card. Trade cards also helped to consolidate and enlarge new national markets for manufactured goods like soft drinks, soaps, and patent medicines. Some of the earliest ads for boxed cereals, canned foods, and candy took the form of trade cards. Buck Duke's American Tobacco, the company that monopolized the cigarette trade with its Bonsack cigarette-rolling machine (see chapter 3), spread the "good news" of its mass-produced cigarette through trade cards and helped to popularize (already in the 1870s and 1880s) what we now know as baseball cards. Indeed, it was cigarette manufacturers that invented them.[50]

The golden age of trade cards extended from 1870 into the early 1900s, by which time magazine advertising was beginning to displace them, coinciding with the rise of mass-distributed periodicals reaching middle-class American homes. Unlike earlier elite magazines, these new media relied on advertising rather than on subscription income for profits. This transformation began in the 1870s, when *Scribner's* and *Harper's* first agreed to publish full-page ads. Advertising agencies founded by N. W. Ayers in 1869 and J. Walter Thompson in 1878 provided manufacturers with professionally designed ad copy, whose placement in popular magazines and newspapers they also helped arrange. When Philadelphia publisher Cyrus Curtis in 1883 set subscription rates for the new *Ladies' Home Journal* at only 50¢ per year, the plan was to attract a much larger readership than had been possible with ad-free magazines — and to deliver customers to eager advertisers. Within five years, Curtis publications (including the *Saturday Evening Post*) had twice the advertising of any competitor, with packaged goods eventually filling most of those ad pages. And so was born our modern bond between mass entertainment media and full-throttle advertising, the branded consumer culture that saturates our world today.[51]

Advertising, along with packaging, became a science about this time. Wharton School advertising professor Herbert Hess understood that the goal of advertising went well beyond introducing consumers to new products; the whole point was to create desire. In one of his early texts (1915), Hess insisted that "each business is directly related to some one of our sense experiences." The manufacturer was to use imagery in advertising (and on packaging) to awaken "desire where people's senses have not yet been aroused to appreciate his particular article." So smell would be aroused in the image of the steaming coffee cup; smoothness would be suggested by "the quivering and translucent gelatin." Visual attraction was especially important: "Each advertisement should consist of an intensity strong enough to force itself into the consciousness" by its bold type, seductive illustration, or striking colors. Like other modern hucksters, Hess stressed the instinctual appeal of certain colors: "A red ground with white or black letters is always one that attracts the eye and tends to cause a halt in movement." But yellow (suggesting light) with red could also attract. Colors could also be used to evoke specific feelings; steel gray suggested durability, for example, while green symbolized water, purple implied royalty, and gold prosperity.[52] Packaging here was key, since the labels on such packaging were, as scholar Charlene Elliott notes, "standing advertisements on the store shelves." Carefully constructed labels functioned essentially as "two-second commercials" and could persuade potential buyers to open their wallets. Packaging was sometimes even said to be more important than what was inside. Commercial artist and product designer W. A. Dwiggins in the 1920s was explicit about this, claiming that the "thing inside the box is of less importance than the box." Bodily response to such ads was a simple matter of human psychophysiology: "digestive organs are trained to exude their juices, not at the sight of food, but at the sight of a rectangular pasteboard prism striped with alternative diagonal bands of white and red."[53]

By the 1920s, the scholarly study of the psychological impact of color, tint, and shade in print advertising and labeling was well underway. Matthew Luckiesh, a lighting engineer at General Electric in Schenectady, New York, claimed that when it came to packaging at least, both sexes preferred blues, reds, and violets (all rare in nature) to greens, yellows, or orange. Richard Franken of New York University and C. B. Larrabee, publisher of the advertising trade journal *Printers' Ink*, concluded that product names should have an appropriate "feeling tone," be easy to pronounce, and sound "good" (Lemon Crush, not Lemon Squash, for example). Much of this "science" appears somewhat arbitrary: blue was said to evoke serenity, for example, but could also be depressing if in the wrong shade or tint. Market surveys in the 1920s found that yellow cans sold the most coffee and that packages in red or black appeared heavier to consumers than those in yellow or blue.[54] Clever ad men were thereby able to elevate a product simply by strategic naming and careful coloring, transforming the utterly plain or undistinguished (think soap or oats) into something quite special—and desired.[55]

Patent medicines were a formidable line of packaged goods, retailed in vast quantities as a result of attractive packaging, decorative labels, and clever ads. Promotional costs for patent medicines typically comprised 30 to 40 percent of sales revenues, with an estimated half of all American advertising dollars (in the 1890s) springing from this one often dubious source. Unregulated by government, patent medicines offered cures for ailments for which trained doctors often had no remedy. Makers of these cure-alls were often in fact physicians—like Dr. David Jayne of Philadelphia who, after 1847, made a fortune selling

**Figure 2.4** A trade card from the 1870s promoting Hires's Root Beer dry concentrate as a temperance drink and patent medicine. Note the early image of the child as advocate of an innovative product. Private collection of Gary Cross.

hair tonics and Sanative Pills (for the liver). Lydia Pinkham's Vegetable Compound, containing 18 percent alcohol as well as roots common in herbal medicines, promised to cure women's complaints and sold quite well, as did Winslow's Syrup for teething babies (the compound contained morphine, hence the quieting effect).[56]

Patent medicines would often display reassuring images of mothers or grandfathers, delighted children, or exotic "natives" in poses suggesting oriental mystery, all of which implied wholesome or extraordinary powers of the particular ointment or emollient in question. And lest we think that advertising of this sort only influenced the gullible or desperate for relief, we should recall that the distinction between patent medicines and ordinary foods and beverages was often blurred in early ads. Trade cards were used for both, and products would often move from being medicinal first to an ordinary consumer staple later. A number of tonics promising a cure for blood and liver ailments or dyspepsia or nervous exhaustion ended up as popular "soft drinks": sarsaparilla and ginger ale are two notable examples.[57] Moxie, one of the earliest soft drinks made in America, was first sold in the 1860s as a patent medicine called "Moxie Nerve Food."[58]

Products as simple as ordinary cereal benefited from clever labeling and imagery. Take, for example, humble rolled oats. In 1854, Ferdinand Schumacher's American Cereal Company introduced high-volume machinery to make rolled oat breakfast cereal, but the founder failed to exploit this advantage with an aggressive sales campaign. A more market-savvy Henry Crowell took over the company in 1877, however, and embellished the rolled oats package with the image of a wholesome Quaker, sold now in cardboard cartons rather than straight and loose out of barrels. The decorated box became an advertisement in itself, and rolled oats for many Americans became simply Quaker oats. The image of the honest Quaker had long been a familiar symbol of rectitude, and one used in patent medicine plugs. But Crowell introduced a new and thereafter widely adopted marketing principle. Instead of employing travelling salesmen to win over retailers, Crowell with Quaker oats used advertising and packaging to appeal directly to consumers. And Crowell's success was quickly imitated. Advertising and package labeling in the 1880s and 1890s shifted dramatically toward simple, eye-catching images, following (if not leading) a kind of modernist aesthetic, encouraging customers to form an emotional bond with the product. Some companies used romantic images to reestablish a relationship lost with the decline of the neighborhood grocer. Chris Rutt, a newspaperman who bought the Pearl (flour) Milling Company, found that he could profit by labeling his otherwise unexceptional pancake flour with an image drawn from Baker & Farrell's popular minstrel act. Performers (black or white) in such shows would dress up in blackface and sing a song about Aunt Jemima, a happy-go-lucky black "mammy." The pancake flour associated with Aunt Jemima evoked a feeling of "southern hospitality."[59]

Similar lessons were learned and put into practice by Campbell's

Soup, a Philadelphia-based company founded in 1869. While already a leader in manufacturing cans and filling machines, Campbell's really only took off when the nephew of its founder, John Torrance, reduced the water content of the company's prepared soup in 1898, producing a concoction that homemakers could serve by simply adding water and heating. Torrance was also early to recognize the advantages of becoming a national brand distinguished by a simple and unified set of branding imagery. And so rather than create a different look for each type of soup, Campbell's went for a visually appealing but simple red and white can adorned with a picture of the medal won by the soup in a Paris exhibition held in 1900. Campbell's ads also stressed ease and freshness, selling Americans on the idea that soup could be the centerpiece of a family meal. The label announced that the contents were of Campbell's quality and that consumers should "accept no substitute." The appeal was reinforced with massive ad campaigns on trolleys and billboards and in newspapers and magazines, including middle-class magazines like *Saturday Evening Post* and *American Magazine*. The results were striking: Campbell's output in 1898 was only half a million cans; by 1924 it had risen to eighteen million.[60]

Scotch tape, Kleenex tissues, Hershey's chocolate bars, and Crisco shortening all managed to dominate their respective markets with forceful trademark advertising. Wrigley's Spearmint Gum had comparable success with its green "spear" package design (1905). But the trademarked image did more than expand sales. Business historian Richard Tedlow has commented on how the trademarked brand created customer loyalty, insulating the manufacturers of that brand (to a certain extent) from price competition. Poor contents of course could always tarnish any brand, but with satisfactory quality and a provocative trademark, a respected brand name created a belief in the superiority (or reliability) of a product that was often no better than its no-name (or poorly named) competitors.[61]

One of the best examples is Coca-Cola. There may well have been something special about the secret formula that is so closely guarded at the Atlanta headquarters, but early Coke certainly varied in its "mysterious mix of citrus oils and kola (caffeine)." One key to the drink's success, though, was its distinctive, trademark bottle. Advertising worked

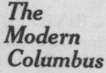

Figure 2.5 Images of youthful vitality and patriotism enhance this ad's effort to make Campbell's tomato soup a daily indulgence. *Saint Nicholas Magazine*, February 1908, 30.

with the package to make it "impossible for the consumer to escape Coca-Cola," as Harrison Jones of Coke's sales office bragged in 1923. Advertising was in fact a major part of Coca-Cola's budget from the beginning of the brand in 1886 (see also chapter 4). In 1892, advertising accounted for 20 percent of Coke's expenditures. In 1913 alone, the Coca-Cola Company distributed 5 million lithographed signs, 1 million calendars, 10 million matchbooks, 50 million paper doilies, two million soda fountain trays, 50,000 window trims, 69,000 fountain displays, and 200,000 signs for refreshment stand walls, all featuring the Coke trademark. So valuable had the brand name become that the company had to wage thousands of trademark infringement cases against copy-cat colas by 1926. And courts were by and large accommodating. In 1912, a judge held that cola manufacturers who used "coca" or even "kola" in their labels were infringing on the rights of the Atlanta juggernaut.[62]

## Winning Over Skeptics

Despite the appeal of ads and packaging, Americans did not always trust packaged goods—especially goods preserved in cans. Manufacturers sensed this reluctance, of course, and appealed to governments for (limited) regulation to consolidate markets. Large packers like Armour and Swift campaigned for a federal meat inspection law in 1891, for example, hoping to win the trust of distant markets while also driving local butchers out of business. Name-brand manufacturers similarly supported legislation to push out cheap manufacturers of "adulterated" goods and to preserve the integrity of their own packaged and bottled goods. Culminating this drive was the 1906 Pure Food and Drug Act, which established what would later become our modern FDA.[63]

This residue of distrust may explain why the advertising of packaged or canned goods often went beyond selling a specific product. Ads tried to convince consumers that packaged foods were pure and safe (even if the competition's goods were not) and, even more, that packaged foods offered advantages not seen in bulk or fresh food. Canned or packed foods were advertised as safer or more "sanitary" than unpackaged or even fresh foods. Appeals such as these were part of what the historian Jackson Lears has referred to as "the modernist quest" for hygiene and efficiency. At the same time, ads went further trying to transform habit and desire with packaged goods.[64]

In the early years of the twentieth century, Nabisco regularly touted the superiority of its packaged Uneeda Biscuits over the traditional cracker barrel. "When you open a package you find them so oven fresh that they almost snap between your fingers," thanks to the moisture-proof package. Uneeda Biscuits were wrapped in waxed paper and packaged in colorful pasteboard emblazoned with a label depicting a charming boy dressed in a yellow raincoat, all to assure customers that Uneeda Biscuits were crisp and tasty, by contrast with moisture-tainted crackers from the common cracker barrel. A 1913 ad for the packaged foods industry in *Collier's* made this same point: "Back when Tabby stuck his inquisitive nose or her soft velvet paws into the butter tub or . . . uncovered sugar barrel," we didn't know about "germs and microbes as we do today." And no one wanted sugar handled by the grocer just after "patting his cat, combing his horse." The solution was modern factory packaging—even if consumers had no idea what conditions in the factory might actually be like.[65]

Similar themes were repeated by apologists for canning—like John Lee, a successful canned goods jobber from California in the 1920s. Canning was superior to fresh food because it eliminated exposure of fruits and vegetables "to the air or to dust, odors and decay" during shipping from distant farms. For Lee, mechanized processing meant clean and wholesome food, especially since the only processing was through the "natural sterilizer" of heat. Canned vegetables were just cooked vegetables—like those cooked fresh at home—and inexpensive because they were canned when and where they were plentiful. Lee waxed enthusiastic: "Canning is God's method of food economy for a world's people whose numbers are rapidly increasing." Canning made possible the safe and cheap transport of "the surplus of Nature's bounty . . . to any part of the world" and put "summer products into the winter pantry and the products of winter upon the summer tea table."[66]

As if to reassure a concerned public, major-brand companies regularly insisted that their "kitchens" were spotless. Schlitz beer of Milwaukee thus bragged in 1904 that while American housewives washed their tableware only once, Schlitz washed their bottles four times. Ordinary mortals used city water, but Schlitz's came from wells fourteen hundred feet below the surface, with the brew itself filtered through "white wood pulp," guaranteeing a "healthful pure beverage."[67] This appeal to beer as "healthful" at a time of looming Prohibition (1919–1933) certainly went beyond addressing claims of safety. The makers of Welch's grape juice likewise insisted that the company used no artificial coloring, chemicals, or added sugar—or alcohol—to reassure a public fearful of unscrupulous packers and processors.[68]

And some major-brand food processors advocated for even more stringent vigilance. In 1913, with the support of the Westfield, Massachusetts, Board of Health, companies as diverse as Kellogg's (cereal), Crisco (shortening), Moxie (the carbonated drink), Knox (gelatin), Beech-Nut (peanut butter), Karo (corn syrup), and Rumford (baking powder) demanded the elimination from all packaged foods of all unnatural additives—including coal tar dyes, alum (used in pickles), apple stock fillers (used in jams) and benzoate of soda.[69]

But the argument for packaging went beyond purity and protection. It also laid claim to another key virtue of modernity—convenience and time saving. Van Camp's beans claimed to save consumers sixteen hours of toil, compared to the time required for home-preparations. No longer would the homemaker have to endure the misery of a hot August kitchen cooking beans. Instead, canned pork and beans could be served after a few minutes of heating in a sauce pan. Heinz offered to take over all of the family's tasks of home preserving, assuring that skilled workers could seed the cherries and cook the choicest fruits to just the right temperature and for just the right length of time. And even a child could prepare the gelatinous dessert of Jell-O.[70]

Not just foods but other packaged goods promised ease and convenience. Colgate's tube of dental cream flowed neatly onto the toothbrush; and Gillette's safety razor was like "taking the elevator and not the stairs" by delivering a shave at home without the bother of old fashioned razors—which always needed resharpening. *Collier's* promoted efficiency and time saving even in its Five Foot Shelf of Books, consisting of distilled selections from the classics chosen by Harvard's eminent Charles Eliot. These carefully preselected works spared the reader wasted time and effort with "useless, fast dying books." Targeting aspiring business and professional people lacking time or money for college but still eager for an education, the Five Foot Shelf reduced the task to something like taking a pill: fifteen-minute readings per day were offered with readable guides "that truly picture the progress of civilization." These works of literature, science, and the arts were supposed to prepare busy men to "think clearly" and "talk well," helping them to join "the successful men of today."[71]

If Eliot and *Collier's* could offer predigested packets of serviceable knowledge, it should perhaps come as no surprise that food producers could promise personal improvement via packaged cereals or canned beans. Quaker oats, Puffed Wheat, and Puffed Rice were routinely touted as miracle foods. The technology used was admittedly impressive: grains of rice or wheat were rapidly heated in pipes, turning their inherent moisture into steam. The hard grains would then explode into soft and easily digestible "puffs" in a process akin to making popcorn. Four times more porous than bread, puffed grain was advertised as breaking up starch granules in ways that simple home cooking could not. Van Camp's steam-cooked beans likewise delivered a "nut-like" flavor permeating "every atom with a delicious taste," while simultaneously making beans "finally digestible."[72]

Advertisers also claimed that canned and packaged goods transcended nature's limitations. Campbell's bragged that one of its historic obligations was to "train nature to do her perfect work," as when the company crossbred tomatoes for its Camden, New Jersey, soup works. Borden's condensed milk claimed to have solved the "infant feeding problem" because its canned milk was "sterile and safe," and in this respect superior even to milk fresh from the cow. And Quaker oats spared no hyperbole in singing the praises of its puffed cereals, said to be crisp and yet delicate, serviceable both as a "food and confection" (since boys ate them like peanuts and girls ate them like candy).[73]

Equality and ubiquity of access were commonly stressed virtues of such foods. No matter where one lived, "even in the hot, stuffy crowded city, you have the advantage of our fresh green fields and fertile fields when it comes to Campbell's vegetable soup."[74] Over and again, makers of ketchup, canned fruits, and vegetables insisted that they had mastered the art of capturing or even improving on nature,[75] while also inviting consumers to embrace new dietary habits. Makers of Cream of Wheat, the breakfast cereal, claimed that this could be served three times a day, even as a dessert. Horlick's Malted Milk (in powder form) was not only "invigorating" when drunk anytime during a tiring day, it could also make a quick lunch for the busy child or adult. Consumed in dried concentrated form as a Lunch Tablet, it could be taken anytime, anywhere. Snacking had not yet received the legitimacy it would later have—we know from ngrams that the word isn't in common use until the 1950s—but times were clearly changing.[76]

In this respect, the push to consume ever-greater portions of packaged foods, especially those "intensified" with sugar and fat not found in nature, threatened time-honored values of moderation and restraint. But to close the deal, manufacturers often resorted to more subtle and indirect appeals, including appeals to the delighted and desirous child. With roots in the days of the trade card, manufacturers decorated their varied pitches for Hoyt's German cologne or Star cough drops with images of spunky boys tipping rowboats or girls enjoying a tree swing. These images made light of "naughty" children's behavior and even greed (as when two small girls were shown fighting over a piece of taffy). In many ways this was simply an early exploitation of a broader cultural trend: adult fascination with a kind of "wondrous innocence" and "the cute" that emerged after 1880 and culminated in the comic strips, ads, and cover art of popular magazines in the subsequent century. But the image of the delighted (if slightly naughty) child also became a stand-in for a public embrace of "natural desire," part of a long-emerging rejection of Puritan self-denial and utilitarian values. A typical example is the 1905 image of the "Campbell's Kids," two pudgy, red-apple-cheeked youngsters in overalls created by Grace Drayton, adorning a Campbell's soup advertisement first in the 1905 *Ladies' Home Journal*. The Kids became a centerpiece of thousands of ads that helped make Campbell's soups nearly ubiquitous in American pantries, images that also appeared on postcards, buttons, and dolls. Similar images of children at this time were used to sell other goods, in-

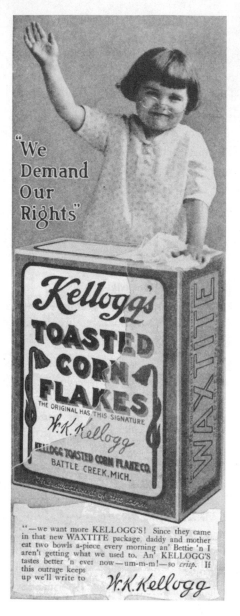

**Figure 2.6** In a massive ad campaign in the eight years after Kellogg's introduced corn flakes in 1906, this was a common image—the desirous child, here with the added "humor" of associating her with the suffrage movement. *Saturday Evening Post*, July 4, 1914, back cover.

cluding Morton's salt, Fairy soap, Clicquot ginger ale, Uneeda biscuits, Fisk tires, Cracker Jack, Armour canned meats, Post Toasties, Dutch Boy paints, Calumet baking powder, and Skippy peanut butter.[77]

Why were kids used to trademark and label such a wide range of products? Some of these images were actually directed at the child consumer. This was certainly the case with the Brownie camera (see chapter 6), but there are other examples. F. E. Ruhlin at the company making Cracker Jack insisted that the Cracker Jack boy (and his dog) were designed to "work its way into" childrens' memories and to "make friends with them."[78] But most of these ads were directed toward adults. The motif was meant to appeal to parents' concerns about the health and development of the growing child, a vital concern at a time when infant and child mortality was still a significant fear.[79] A more dominant theme, however, was the new appeal to the natural right of children and, by implication, adults to partake freely (and ideally, in quantity) in the abundance of American industry. Quaker warned consumers not to make cereal "duty food"; the "creamy flakes of Quaker Oats" would certainly provide your child with energy, but they were also so "delightful" that the uncoerced child would naturally desire them. The implied claim was that there was no way to control a child's desire, and that it was therefore better to help them satisfy than to suppress that desire. A pure, even selfish, desire for consumables was no longer something to control or discourage; commerce in this new form asked consumers to regard unbridled desire (and therefore consumption) as natural or even healthy. This again was part of a long campaign to convince Americans to abandon lingering Puritanism, to persuade them that their longings and desires were on the whole natural, inevitable, and good—and fulfillable through commerce.[80]

## The Take-Off of the Packaged Pleasure

There is something special about the span of time between roughly 1880 and 1910, when a confluence of inventions accelerated trends dating from when pots were first used to move, store, and cook. This rapid expansion of new technologies, especially of bottling, canning, and boxing, laid the foundations for a new kind of consumer culture in the form of the portable personal commodity, accessed cheaply and virtually anytime, anywhere. As we shall see, this packaging and tubing revolution extended beyond the can, bottle, and box to include the paper wrapping of the cigarette, the groove of the phonograph record, the film strip of the camera, and the tracks on roller coasters. Although these things had vastly different medical, social, and psychological impacts, they all share common roots in a cluster of new technologies and marketing innovations arising over the span of a mere generation. Packaged pleasure technologies radically transformed the human "experience economy," be it with new access to an amazing range of foods, drink, and psychoactive drugs, or, as we will see, in the nicotine-delivery system of the modern cigarette and the plethora of superfoods based on sugar, chocolate, and carbonation. And those technologies very often transformed and intensified what it meant to see and to hear or even to move. By compressing sensuality, in effect, these technologies revolutionized what it meant to experience life. Such was the power of the package.

Of course if what emerged from this was novel forms of consumption, it also came with a certain cost. We have already seen how tastes became more uniform, as millions embraced the same brands of soup, cereal, and soda. Items came to be desired as much for their labels and associations as for their flavors and nutritive value. Old taboos disappeared, but so did seasonal, ritual, and festive foods and drinks. Mass commercial containerization also made it harder to know where food actually came from. And with the exotic becoming ordinary, the experience of rarity itself was sometimes dulled. The packaged pleasure may even have made us more hedonistic, with consequences we have not really thought enough about, even today.

In the following five chapters we will explore more fully the diverse manifestations of the packaged pleasure, beginning with the most extreme and morbid form, the cigarette, and proceeding to perhaps the most subtle, the amusement park—each, despite sharing characteristics, distinct in its impact on our lives.

# 3

# The Cigarette Story

It would be hard to name a more consequential—indeed, more deadly—example of the packaged pleasure. Cigarettes are the chief means by which nicotine is now taken into the body; that is why most people smoke, and have always smoked: to partake of the diabolic delights of nicotine pharmacology, to which they often become addicted. Cigarette manufacturers call nicotine the *sine qua non* of smoke,[1] and cigarettes have become the most widely used (and abused) drug on the planet—by far. Six trillion cigarettes are smoked every year, enough to circle the globe some fifteen thousand times or to stretch from the earth to the sun and back, with enough left over for a couple of round trips to Mars.

How, though, did this quintessential packaged pleasure come into being?

Tobacco use has a long history. American Indians are known to have burned and consumed several different species of the tobacco plant (*Nicotiana tabacum*, *rustica*, and so forth); tobacco was grown, dried, and smoked in ways quite different from the smokes of our own industrial age. Like other forms of sensual intensity, tobacco was commonly used for ritualistic purposes—to seal an agreement or in religious ceremonies of various sorts—and there was little of the compulsive, intensive, self-destructive use we find so often in our own times. The cigarettes smoked today are quite different from those smoked prior to the nineteenth century, and it was really not until the rise of mechanized processing and rolling at the end of that century that smoking was transformed from a ritual or recreational drug into a mass addiction. The product itself was different, as was the means by which it was produced and marketed. And the scale on which it was—and is—used.

Some index of this transformation can be seen in the fact that Americans in 2014 will smoke about 300 billion cigarettes; contrast this with the mere 2 or 3 billion smoked as recently as 1900. We like to think of tobacco use as something from our distant cultural past—and Hollywood and shows like *Mad Men* reinforce this myth—but the fact is that Americans today consume far more tobacco than in previous centuries. In 1800, for example, Americans consumed only about 100 million pounds of tobacco.[2] By 1860 this had increased to 300 million pounds, and by 1910, to 1 billion pounds. Production would not peak until the decades after the Second World War, when American factories would process about 2 billion pounds of tobacco every year. Elsewhere in the world the rise has been as spectacular or even more so: the Chinese smoked only about 7.5 billion cigarettes in 1911, for example, vs. nearly 2.4 *trillion* in 2012.

The pattern of use has also changed dramatically, in consequence of mechanized manufacturing, mass marketing, and the use of new forms of curing that created a more addictive—and deadly—tobacco product. A typical smoker of cigarettes today will take two hundred or even four hundred or more separate "hits" (puffs) in the course of a single day, repeating this self-dosing ("titration") day in and day out, without ever missing a day. Few regular cigarette smokers ever take a day off. But we can hardly imagine such an intense, routinized, and compulsive use prior to the invention of the modern cigarette. Tobacco was just one of many different psychoactive substances occasionally indulged in; use by Native Americans, for example, seems to have had more of a ritualistic than a routine or compulsive character; there may well have been some addicts, but that cannot have been very common.[3]

The irony here is that cigarettes, ultimately so destructive, when first introduced in a commercial form in the nineteenth century were commonly regarded as a milder, easier, and more convenient way to smoke, compared with pipe tobacco and cigars. Cigarettes were less harsh and less "difficult" to smoke; that was part of the attraction, and early on the fear, that as a cheaper, milder, and indeed "faster smoke" (like fast food) the cigarette would attract women and children. Cigarettes were also seen as a kind of sissy smoke for dandies and overly refined Yankees, with again this irony that while commonly regarded as "milder" and therefore less harmful than other forms of tobacco use, the reality was that by virtue of being routinely inhaled — indeed, you *had* to inhale to obtain "full satisfaction" — cigarettes were actually far more deadly than other forms of tobacco, and far more addictive. They would also become a far more carefully — and craftily — engineered consumed product, with literally tens of billions of dollars spent by manufacturers (just in the United States) on cigarette design.

Cigarettes are most often traced to the nineteenth century, though the origin can be stretched much further back, depending on how we choose to define our terms. If by "cigarette" we simply mean a "smallish cigar" — as the French etymology suggests — then cigarettes were clearly found in pre-Columbian America. Native Americans imbibed the smoke from burning *Nicotiana* in many different forms and through several different orifices (mouth, nose, and anus) and some of these small smoke generators we could certainly call thin, short cigars. Hundreds of images of tobacco use are preserved on Mesoamerican pottery and in Mayan codices, and in some of these we clearly see the smoking of smallish cigars. If by "cigarette," however, we mean smokeable tobacco wrapped in paper, then the cigarette has a more recent origin. Spanish beggar boys from the seventeenth century are known to have gathered up tobacco scraps and rolled them into tubes made from newspaper waste, and these were smoked. As the first known examples of tobacco being smoked in paper, these poor-boy *papalets* could also be dubbed the first "cigarettes."[4]

More conventional histories trace the modern cigarette to nineteenth-century events, most of which involve warfare. One early form of cigarette use — meaning tobacco smoked in paper — took place in 1832 during the Siege of Acre (in Palestine), when an Egyptian cannoneer tried rolling some of the tobacco he'd been given (as a reward for rapid firing) in the paper he'd been using to roll his gunpowder. (Some versions of the story say their pipes had been destroyed.) The paper-rolled tobacco seems to have provided a suitable smoke, and voila, the cigarette was born. Cigarettes were given a further boost during the Crimean War of 1853–1856, when British soldiers copied their Ottoman and Russian counterparts, who had recently become accustomed to rolling tobacco in newsprint. These eastern, one could say Orientalist, origins of the cigarette habit are recalled in the names of countless early cigarette brands, smokes with names such as Murad, Abdulla, Mecca, Omar, Camel, and so forth.

Cigarettes were a trivial form of tobacco use for most of the nineteenth century, however. As late as 1890, cigarettes constituted only a tiny fraction of the entire American tobacco trade; tobacco was far more often chewed, rolled into cigars, or smoked in pipes. Cigarettes were smoked by dandies and the effete rich and errant youth, and broader use was hampered by moral indignation and, for a time at least, the enactment of bans on the sale of cigarettes: from 1890 to 1927 some fifteen U.S. states prohibited the sale of cigarettes, fearing moral corruption of women and youth and bodily harm from the filth so often found in cigarettes.[5] Cigarettes were not yet terribly popular.

## Susini's Honradez Machine of Havana

Mechanization and mass marketing — and the free distribution of cigarettes to soldiers during the First World War — changed this. As with canning, mechanization dramatically lowered the price of cigarettes and prodigiously raised the number that could be produced per unit of time and labor. Cigarettes prior to the 1880s were almost always rolled by hand; the women and girls employed in such factories typically could only roll about four or five cigarettes per minute, and factories employed many hundreds of such girls, each with their own rolling station (see figure 3.1).

Mechanization actually began somewhat earlier than is commonly imagined, and the story actually begins in Cuba. The oldest known automatic cigarette-making machine was developed by Don Luis Su-

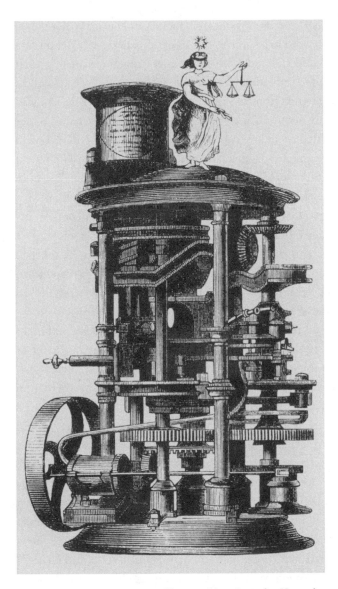

**Figure 3.1** Susini cigarette-rolling machine from the Honradez ("honesty") factory in Havana, Cuba, 1867. Susini's machine could roll sixty cigarettes per minute, essentially by imitating the hand-rolling process. The Susini would be superseded by the Bonsack machine, which could roll cigarettes much faster—and without those little twists at either end.

sini for use in his Honradez ("honesty") factory in Havana in the 1860s. The machine was said to be able to roll some sixty cigarettes per minute (see figure 3.1) and differed from subsequent machines in that each cigarette was rolled individually, with little twists on each end, just like hand-rolled cigarettes from the time. The resulting *cigarros* were also quite short, only about an inch and a half long. The fact that paper twists were added at either end was another notable contrast with machine-made cigarettes from a later era, which, by virtue of being cut from a continuous "rope" of tobacco, had an open-cut "face" at both ends. This open face caused the tobacco to dry out faster than with twisted-end cigarettes, which is why cigarettes made from continuous-process machines required treatment with "humectants" such as glycerine, molasses, and eventually diethylene glycol: to keep the tobacco from drying out. Cigarettes even now contain about 10 percent additives by weight, with nearly half of this being chemicals designed to counter this side effect of continuous process manufacturing.

Today we tend to associate Cuba with cigars, but in the middle of the nineteenth century the island was also a significant producer of cigarettes, with some two million packs (*cajatillas*) exported every year. (The size of these packs was not yet uniform, containing anywhere from ten to fifty sticks each.) Cuba in 1848 boasted around four hundred tobacco factories, several of which were making what we today would call cigarettes, referred to at the time as *cigarros, cigaritos, papelitos, papeletas, papeletes, papelotes,* and *papelillos.* (Cigars, by contrast, were most commonly known to Cubans as *tobaccos.*) Cuba's industry underwent a certain consolidation over time, and by 1861 there were only about thirty-eight *cigarrerias* (cigarette factories) on the island. Cuban cigarettes were also "distinctly different than Russian cigarettes made of Middle Eastern and Turkish tobacco with mouthpieces and filters."[6] Cigarettes from this era were short but also quite thin—only about an eighth of an inch thick—which is why a kind of "roach clip" was often used to grasp the burning rod, which could then be smoked right down to the bitter end.

The first known factory exclusively for making cigarettes was established by Don Luis Susini in downtown Havana in 1853. At his Honradez factory, Susini used tobacco judged inappropriate for cigar manufacture, meaning leaves that couldn't be rolled whole or were broken or otherwise imperfect. Early cigarettes were often difficult to keep lit, which is also why manufacturers cut tobacco leaves into very fine shreds and would eventually add burn accelerants such as sodium or potassium citrate[7] (which is also why cigarettes cause so many fires: they tend to keep burning when dropped). Cigarette makers also had a reputation for using inferior quality ingredients, and Don Luis Susini & Son tried to counter this by giving dignified names to his cigarettes and factory (*La Honradez* means "honesty" or "integrity"). Trademarking at this time carried little legal force, however, and counterfeiting was rampant, which is one reason Susini started printing distinctive labels on his products. His were some of the world's first colored labels, using the newly developed process of color lithography. Susini's colorful labels from the 1850s, '60s, and '70s are some of the most beautiful packaging ever: artfully designed labels were on Susini cigarettes by the 1850s, even before the widespread dissemination of trade cards in the United States. Other means were used to discourage counterfeiting, including regular changes in package art and brand imagery. Tony Hyman in his history of the Honradez factory points out that scrapbooks preserved in the Jose Marti National Library in Havana contain nearly four thousand different labels from early Cuban cigarette factories.[8]

Cigarette packaging at this time was technologically progressive. Color lithography was actually first used commercially in the cigarette business—one of many marketing innovations launched by cigarette manufacturers. (Others much later include skywriting, skycasting, billboard photolithography, coast-to-coast radio sponsorship, stop-motion animation, and so forth.) Color lithography revolutionized packaging and must have been quite astonishing to consumers unaccustomed to such exciting sights. Here is how one visitor described Susini's Royal and Imperial Honradez factory in Havana in the late 1860s:

> In the lithographic, drawing, and engraving room, I saw what I had never seen before in any other establishment, and which, I am told, is an entirely new discovery—the process of drawing on stone by chemical action and machinery. This is the machine known as the "Magneto-Electrique Machine," the invention of Mr. E. Gaiffe, a Frenchman, and which took the

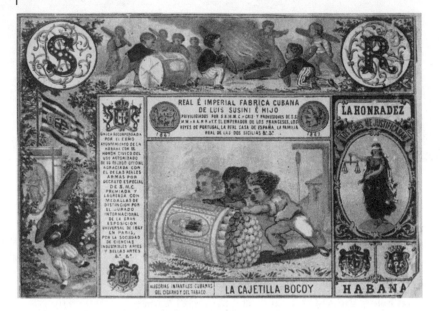

**Figure 3.2** Cigarette packaging label from Luis Susini's Honradez factory in Havana. Susini's factory was technologically sophisticated and the first factory anywhere to introduce colored labels for distinct packaging, beginning in the 1850s. This sample here is from 1867.

prize at the World's Fair and others. This machine, which has for its principal organ electricity, is the first that has been practically put in use in this sort of industry since the days of Franklin, the discoverer of electricity. The principle of the machine rests upon the interruption of the currents by a composition, or isolating ink, of which the design upon the matrix is composed.

By a circular motion of both the surface to be engraved on and the graver, and with the assistance of an electric magnet, pointed with a diamond, a complete and perfect drawing can be made without requiring any human assistance.[9]

Susini's factory was often admired by guests, who marveled at his exploitation of telegraphy, pneumatic tubes for interoffice communication, and new and improved methods of packaging. Susini's machines also gave rise to an entirely new rhetoric of consumerist gigantism.

Count Joseph de Susini-Ruiseco in 1872 figured that with cigarettes being about 70 millimeters each, the 294 billion produced every year globally (by his exaggerated guesstimate) were altogether 20,600,000 kilometers in length. And since the earth has a circumference of 40,000 kilometers, this meant a global annual production sufficient to make a chain of cigarettes that could circle the earth 514 times. Or cover a surface of 144 million square meters. Or in 7.5 years, enough to make a continuous cable of smokes stretching from the earth all the way to the sun.[10] Susini-Ruiseco's hope was that mechanization would help reduce the costs required to produce such a vast quantity of cigarettes; mechanization was a point of pride and, when exhibited at the Universal and International Exposition in Lyon in 1872, Susini's machine won a medal. Susini-Ruiseco compared his machines to the telegraph, the typographic press, and the electrographic machines of Gaiffe, all of which were taken as progressive boons for humankind.

Yet another rationale for mechanized production, though, was to improve workers' health. The idea here was that by eliminating direct contact with the product, mechanization would prevent damage to the eyes, lungs, and throats of workers who might otherwise suffer harms from handling or inhaling toxic tobacco dust. One Susini machine was able to do the work of twenty hand-rollers; Susini-Ruiseco for this reason defended mechanization as rendering an "immense philanthropic service" to humankind — even if it also took away jobs and livelihoods.[11]

## Mechanization and Mass Marketing

Susini machines were apparently quite finicky and never widely used. The machine was displayed at the 1867 Paris Exhibition and then again at Lyon in 1872, but we should not exaggerate the significance of automated rolling in these early years. As late as 1900, machine-made cigarettes still accounted for only 4 or 5 percent of all U.S. tobacco consumed (by weight); tobacco was far more often taken in the form of chewing tobacco, pipes, and cigars. Cigarettes would not come to dominate global tobacco culture until the 1920s and '30s.

And mechanization was not the only force fostering consumption. Matches also helped popularize cigarettes: phosphorous matches date only from the 1820s, while "safety matches"—which require a special striking surface—date only from the 1840s. Matches helped foster the growth of the cigarette habit by increasing the ease and convenience of making fire, allowing a quick and calibrated ignition. Matches commercialized—we could say radically simplified and packaged—the age-old capacity of humans to create fire while also dramatically reducing the skill required. Liquid fuel lighters did much the same toward the end of the nineteenth century: wind-protected "lighters" were patented in the 1890s and enjoyed a prodigious popularity with the invention of the Zippo lighter in the 1930s.[12]

As with other packaged pleasures, mass marketing was also pivotal in the rise of the modern cigarette. Marketing had begun prior even to the invention of rolling machines, notably with the introduction of colorful cigarette labels printed by the novel method of color lithography. Manufacturers also started placing cardboard stiffeners into cigarette packs, originally to keep the soft packs from being crushed, and soon thereafter realized that these could be used as yet another advertising platform. Cigarette cards were born from this basic concept, and by the 1870s cigarette makers were including images of sports heroes, landscapes, and (eventually) celebrities on cigarette "cards," which, like the trade cards discussed in chapter 2, quickly became popular items for collectors while simultaneously helping to establish the phenomenon of celebrity (and the hobby of collecting; recall that cigarette makers invented what we now know as baseball cards). Marketing in this sense was the perfect complement to mechanization; mechanization increased the *supply* of cigarettes, while marketing helped dispose of that surplus. Marketers produced the "itch" that manufacturers could then step in to scratch.[13]

Marketing became even more important when mechanization dramatically lowered the costs of cigarette manufacturing—by about a factor of ten.[14] Recall that even an expert hand-roller could roll only about 4 or 5 per minute, meaning that with breaks for lunch and so forth, an energetic "girl" working a normal shift might produce only 1,000 cigarettes per day. Mechanization meant that fewer workers could make far more cigarettes. To put this into perspective, recall that by the 1980s, when consumption in the United States reached its peak,

Americans were smoking over 630 *billion* cigarettes per year. If all of these had been rolled by hand, then cigarette making would have been the biggest industry in the country, by far. If hand-rollers rolled 1,000 cigarettes per day and 200,000 per year, this would mean an army of over 3 million rollers working full time to crank out just the cigarettes smoked by Americans. Labor productivity today exceeds that of the nineteenth century by a couple orders of magnitude: 10,000 cigarette workers now produce about a billion cigarettes per day; that's 100,000 cigarettes per worker per day, a hundredfold increase over the productivity of nineteenth-century hand-rollers.

Mechanization of cigarette production was not a single-stage process. Shredding machines were patented in 1866, and other kinds of equipment had been devised to facilitate the preparation of tobacco suitable for rolling. The single most important breakthrough, however, was the recognition that mechanization could be used not just to imitate hand-rolling (à la Susini), but rather to generate a cigarette of essentially infinite length, which could then be rapidly cut into cigarette-sized segments. Here, as in so many efforts to mechanize labor, breakthroughs came only after inventors abandoned the effort to imitate hand movements. James Albert Bonsack of Virginia was the first to develop such a machine, which revolutionized the cigarette trade by introducing what we now know as continuous process manufacturing.[15]

Of course Bonsack was not operating in a vacuum. Cigarette makers had already begun looking for faster and more reliable methods of automation, and the Allen & Ginter Company of Richmond, Virginia, in 1875 had actually offered a prize of $75,000 for anyone who could come up with a reliable machine for rolling cigarettes—meaning one that could dramatically surpass Susini's. In 1880 Bonsack, then only 20 years old, took up the challenge, leaving school to build such a machine from parts in his family's wool mill. The machine for which he was granted a patent in 1881 was an amazing ensemble of rollers, gears, and belts—with the capacity to produce some 20,000 cigarettes in only 10 hours, revolutionizing the nascent cigarette trade and setting into motion a health catastrophe without parallel in all of human history.[16]

Bonsack's machine was different from previous devices in several important respects. Speed was the principal virtue, due mainly to the fact that cigarettes could now be fabricated in an endless stream, much as rope or extruded metal wire was coming to be produced. So whereas Susini's machine had essentially mimicked the hand-rolling process, yielding individually-wrapped cigarettes complete with a little twist at each end, Bonsack's produced a cigarette of infinite length, which could then be cut into appropriate sizes by powerful whirling shears. Key to this new approach was a means by which chopped tobacco could be rapidly funneled through a series of compressing tubes (reprising the wide-ranging historical significance of the tube), yielding a fast-moving "rope" of compressed leaf, absent only the surrounding paper. A large, disk-like spool or bobbin of paper was made to wrap around this fast-moving rope, which could then be sealed with a thin stream of glue. Precision control of feed and cutting speeds allowed an operator to control the resulting cigarette's length and weight, facilitating also the development of standard-sized packs and cartons. Standardized production also allowed the solidification of legal definitions of cigarettes (notably for import and taxation purposes) and helped make possible the sale of standardized cigarette packs from coin-operated vending machines.[17]

Early cigarette-making machines were delicate and could be finicky—and even Bonsack's well-built device required tinkering and adjustments in its early years. Seven Bonsack machines had been installed in U.S. factories by 1884, with an equal number reaching Europe.[18] James Buchanan Duke at W. Duke Sons & Company saw a gold mine in the new technology, and obtained an exclusive contract for future machines. By 1886, Duke had fifteen Bonsacks and by 1889 he had twenty-four, allowing him to crank out some seven hundred fifty million cigarettes per year. With a virtual monopoly on mechanization and a huge stream of profits from sizable margins, Duke's American Tobacco Company—formed in 1890 from a merger with four of his biggest competitors—quickly came to dominate global production. His "Tobacco Trust" continued to swallow up smaller manufacturers until 1911, when by exercise of the Sherman Antitrust Act it was broken up into the separate oligopoly of R. J. Reynolds, Lorillard, Liggett & Myers, and a new (but diminished) American Tobacco Company.[19] (Philip Morris at this time was not yet even making cigarettes, and as

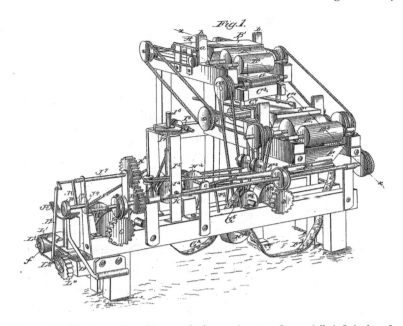

**Figure 3.3** The Bonsack machine cranked out a cigarette of essentially infinite length, which was then cut to size by rapidly spinning shears. Thanks to the Bonsack machine, cigarettes became one of the first consumer "goods" produced according to the principles of continuous processing. The consequence was a dramatic fall in the price of cigarettes, changing them from luxuries into products of ordinary and affordable consumption. U.S. Patent Office, 1881 (no. 238,640).

late as the 1950s still had only about 5 percent of the American tobacco trade. Freebasing chemistry and clever marketing would later propel this cowboy brand into the world's most popular cigarette.)

History does not end with the Bonsack machine, however, and the oft-overlooked fact is that cigarette manufacturers have kept developing ever-faster machines in the century and a half since (see the table, below). Jose Molins started making hand-rolled cigars and cigarettes in Havana in 1874, for example, and by the time his sons Harold and Walter were running the business in London in 1911, the company had branched out into making machines for a wide variety of packaged goods, from cigarettes to tea. Recognizing the value of high-speed production, Molins developed his Mark 1 cigarette making machine, for which a patent was obtained in 1924. In 1931 the company opened a factory in Richmond, Virginia, the center of American tobacco manufacturing, supplying American Tobacco and other local firms with its machines. Walter's son Desmond in 1937 patented the hinge-lid pack, a key element in the 1954 redesign of Marlboro, introduced with the brand's transformation from a woman's to a man's brand. Molins would go on to establish factories in India (Behala, near Calcutta) and Brazil (São Paulo), while also diversifying into other products in the 1970s and '80s (tea-bags, for example, and machines for making corrugated board).[20] Other companies today make cigarette-making machines: G.D (Generate Differences) in Italy manufactures cigarette-making equipment, along with Molins in London and Hauni in Hamburg, but there are not many others. China is now trying to enter this market, as the Middle Kingdom seeks to become the world's cigarette superpower. China already produces 40 percent of the world's cigarettes, and that percentage is likely to continue to grow.[21] The crucial point to keep in mind, though, is that cigarette-making machines are now about a hundred times faster even than Bonsack's first devices.

### Cigarette-rolling machine speeds

| TYPE OF MACHINE (MANUFACTURER) | YEAR | CIGARETTES ROLLED PER MINUTE |
|---|---|---|
| Hand rollers | 1800s | 4–6 |
| Susini | 1867 | 60 |
| Bonsack | 1885 | 210 |
| Bernhard Baron | 1897 | 480 |
| Excelsior III B (Universelle) | 1910 | 250 |
| Gallia UD5 (Decouflé) | 1914 | 500 |
| 4 CC (Du Brul of Cincinnati) | 1919 | 500 |
| Excelsior Rapid (Universelle) | 1922 | 1,100 |
| Mark 1 (Molins) | 1926 | 1,000 |
| Triumph (Universelle) | 1929 | 1,200 |
| Progress Company (Dresden) | 1930 | 1,800 |
| Mark 5 (Molins) | 1930 | 1,000+ |
| Rapid Excelsior (Universelle; improved) | 1930 | 1,300 |
| Decouflé LOB (Arenco) | 1934 | 1,500 |

## Cigarette-rolling machine speeds (continued)

| TYPE OF MACHINE (MANUFACTURER) | YEAR | CIGARETTES ROLLED PER MINUTE |
|---|---|---|
| *Decouflé LOB (Arenco)* | 1934 | 1,500 |
| *Mark 6 (Molins)* | 1935 | 1,000+ |
| *Excelsior Rapid KDC (Hauni-Werke)* | 1949 | 1,350 |
| *Excelsior Super Rapid KDZ (Hauni)* | 1953 | 1,400 |
| *Mark 6 (Molins; improved)* | 1953 | 1,350 |
| *LOD (Decouflé)* | 1955 | 1,500 |
| *LOF (Decouflé)* | 1957 | 2,000 |
| *Mark 8 (Molins)* | 1958 | 1,600 |
| *Garant 1 (Hauni)* | 1959 | 1,500 |
| *Garant 4 (Hauni)* | 1968 | 4,000 |
| *LOG (Decouflé)* | 1970 | 4,000 |
| *Mark 9 (Molins)* | 1971 | 4,000 |
| *Mark 9-5 (Molins)* | 1976 | 5,000 |
| *PROTOS (Hauni)* | 1978 | 6,000 |
| *PROTOS (Hauni; improved)* | 1982 | 7,200 |
| *Mark 10 (Molins)* | 1984 | 8,000 |
| *Mark 10 (Molins; improved)* | 1986 | 9,000 |
| *PROTOS 100 (Hauni)* | 1988 | 10,000 |
| *PROTOS 100 MAX 100 (Hauni)* | 1993 | 12,000 |
| *121P (G.D)* | 2006 | 20,000 |
| *PROTOS M-8 (Hauni)* | 2008 | 19,480 |

Most of these machine speeds are from Hans-Dietrich Klopfer, "The Quest for Speed," *Tobacco International*, October 15, 1992, 29–32, http://legacy.library.ucsf.edu/tid/onj79b00.

Cigarette packaging of course has had to keep pace with these ever-faster rolling machines: there was no point in fast rolling if you couldn't fast pack. Output from the earliest cigarette-making machines was packed by hand but this, too, was quickly mechanized, along with the insertion of packs into cartons. Cigarette making machines today are seamlessly tied to packaging machines, including machines that insert cigarettes into packs, packs into cartons, cartons into cases, and cases into containers.[22] All of this has dramatically lowered labor costs and the time required to crank out cigarettes.

The industry measures this increased productivity in terms of cigarettes per labor hour (CPLH), a measure introduced in the American tobacco trade in the 1970s. In 1973, for example, Philip Morris factories had a composite CPLH of about 9,700. By 1986, this had grown to 16,000.[23] CPLH continued to increase in subsequent years, and by 1990, R. J. Reynolds in the United States had a CPLH exceeding 22,000, with Philip Morris shortly thereafter achieving 28,000.[24] Mechanization by the end of the millennium was allowing R. J. Reynolds's factory in Turkey to produce 2,847,000 packs of cigarettes per employee per year.

The fastest machines in the world now make about 20,000 cigarettes per minute, or 340 per second. Machine expertise has allowed a number of cigarette manufacturers to branch out into other domains. In 1991, for example, Japan Tobacco (a state monopoly) used its expertise in this area to create the MH-5000, "a machine used in the high-speed application of semiconductors on circuit boards." Developed in a joint project with Toshiba Corporation, the MH-5000 can apply semiconductors at the rate of eight per second, the fastest in the world.[25] The more important consequence, however, has been the maintenance of high profit margins, and thus high returns on stock prices. An investment of $10,000 in Philip Morris in 1958, for example, would be worth about $50 million today. Mechanization also means that far fewer workers can produce far more cigarettes: so even though Americans still produce about 400 billion cigarettes per year, these are made by a labor force of only about 10,000. More than one in four of these cigarettes are for export, which makes American cigarette manufacturers responsible for over a hundred thousand annual deaths on foreign shores.

## A Digression on Paper

A cigarette is basically a small cigar in a tight paper bag or sleeve. (The industry calls it a "wrapper," making it sound more like a kind of Christmas present.) If the bag comparison seems strange, that is perhaps because it is easy to forget that no one has ever smoked a cigarette without also smoking the paper that encases it. Cigarettes are typically about 5 percent paper by weight — or .05 grams per 1-gram cigarette — which means that in the twentieth century alone, during which 100 trillion cigarettes were consumed, smokers burned and inhaled about 100 million metric tons of tobacco but also about five million metric tons of cigarette paper. Smokers in the twentieth century inhaled smoke from millions of tons of paper, millions of tons of chemical additives, thousands of tons of pesticides, and much else as well — all from cigarettes sold in attractive, easy-to-access packs.

Early critics of cigarettes were quick to pounce on this oddity that cigarettes seemed to be worse for you even than cigars, with the culprit perhaps lying in the ivory paper sheath. There is an interesting irony here, since cigarettes were often regarded as being "milder" than other forms of tobacco and therefore more dangerous, insofar as this "mildness" made them more appealing to women, children, and other vulnerable innocents. There was also the fear that cigarettes had the power to corrupt the body in novel ways, with the paper being a logical suspect (along with inhalation, since the smoke from cigarettes was more commonly taken into the lungs). Henry Ford in his 1914 "Case Against the Little White Slaver" (with an introduction by Thomas A. Edison) railed against the cigarette on several grounds, chief among these being that your typical criminal was "an inveterate cigarette smoker" with a tendency to hang out in pool rooms and saloons. The physical damage from cigarettes, though, as both he and Edison were convinced, came "principally from the burning paper wrapper."[26] Edison and others suspected that something must distinguish the (deadly) cigarette from the (relatively harmless) cigar, and suspicions fell on the paper and the diverse chemicals with which it had been treated (arsenic, for example).

Lots of attention was given to this cigarette paper question in subsequent decades, especially the 1950s, when the broad confluence of epidemiology, animal experiments, chemical analytics, and studies of human cellular tissues at autopsy showed that cigarettes were behind the nation's dramatic upsurge in lung cancer. One theory was that it was the paper, and not the tobacco per se, that was causing all this cancer. Cigarette paper, upon burning, was known to produce acrolein, a lacrimal agent (makes you cry) used by the French in the First

World War, and the paper-blame theory seemed to fit with the observation that cigars did not seem to be causing much lung cancer (though former President Ulysses S. Grant's death from throat cancer in 1885 was widely traced to his fondness for cigars). The American Tobacco Company in 1953 financed a series of secret studies to determine what was causing all this cancer, and found that it was the tobacco, not the paper. This must have come as a great relief to the Ecusta Paper Corporation, which had been contracted to do the study; it also meant that the world's largest manufacturer of cigarette paper could now profit from the manufacture and sale of cigarette filters, turning tobacco's lemons into its own lemonade.[27]

## Back to Cigarettes—and Flue Curing and Mass Marketing

Many of the categories we associate with cigarette constituents are reifications: "tar and nicotine," for example, which the industry has pushed for so many years as if these were natural and distinct entities, when the reality is that we might as well talk about "tar and cyanide," or "phenol and nicotine," or "nicotine and benzpyrene." Tar is just condensed smoke, and nicotine gets separate billing only to distract from the fact that the alkaloid is only kept in the leaf to create and sustain addiction. Crucial here also is the fact that the cigarette itself is plastic as a defined object. What, after all, is the *sine qua non* of a cigarette? The smallish size? Or the paper?' Or the inhalability of the smoke? Inhalability is arguably the chief defining feature of the modern cigarette; it is equally the fatal flaw, since inhalation of the resulting smoke is what makes cigarettes so deadly and addictive. It is not widely known, but cigarette manufacturers could easily make cigarettes that would not be routinely inhaled—they would only have to raise the pH of the smoke. Cigarettes in this sense are not inherently dangerous but rather dangerous by design. This medico-forensic fact has historical roots and can only be understood by understanding how the tobacco leaf is cured.

Early cigarettes were usually made from inferior tobacco, meaning leaves with holes or tears or some other fault making them unsuitable for cigars. There was not yet any real difference between the leaves used for cigars and cigarettes, not in the Americas at least. In Cuba, both were made from a dark, alkaline, low-sugar (from long air curing) leaf known as *al cuadrado*—by contrast with the lighter, milder, "flue cured" tobaccos or their smaller-leafed Turkish counterparts. Tobacco chemists later realized that the use of these low-sugar tobaccos made the cigarette smoke quite potent and harsh, which also meant that the smoke from these earliest cigarettes was less often inhaled and therefore less likely to cause harm to the lungs.

What changed to make cigarettes far more deadly—and addictive— was the widespread use of flue curing, a technique by which tobacco leaves are cured by heat piped through low brick chimneys (flues), exposing the tobacco to much higher temperatures than in air curing methods. The older technique of air curing had basically involved just drying tobacco leaves in some cool dry place; that was the traditional practice of Native Americans, though by the nineteenth century tobacco was also being cured in barns, with wood fires sometimes used to speed the process. The virtues of high-temperature flue curing were discovered by accident in the 1830s, when a slave on the farm of Abisha Slade in North Carolina was tending the fire in a barn when he fell asleep, awakening later to find a dying fire in need of fuel. Finding no wood, the man gathered up some charcoal from the nearby iron furnace and threw this onto the fire. The extra heat (charcoal burns hotter) turned the leaves a golden yellow and caused the resulting smoke to be exceptionally mild. The heat-treated leaves fetched a significantly higher price at auction, thanks to a new kind of chemistry.

Flue curing as we now know prevents decomposition of sugars in the leaves, meaning that tobaccos could be produced with a high sugar content in the finished leaf. When burned, high-sugar tobacco produces a less alkaline smoke with a pH of around 6, which is easier to inhale than the smoke from air-cured tobacco, which yields a harsh smoke with a pH of 8 or even higher. The milder smoke from flue-cured leaf was easier to inhale and, since it didn't have the free-based "kick" of high-pH tobacco, actually *had to be inhaled* to get full nicotine "satisfaction." The net effect: cigarettes made from flue-cured tobacco were very often inhaled. Inhalation was also a far more effective means of delivering nicotine to the blood and brain (recall that your lungs have an

internal surface area the size of a tennis court), which also made cigarettes far more addictive than any previous form of tobacco use.[28] The
shift from air-cured to flue-cured tobacco was something like the shift
from opium to heroin or from eating to injecting drugs by means of a
hypodermic needle: a new form of "consumption" based on a new and
intensified form of packaged delivery.

Sugar was thus crucial to the rise of the modern cigarette in two different ways. Flue curing preserved a great deal of natural plant sugar
in the cured tobacco leaf (around 20 percent, versus the 2 percent left
after air curing), but sugars were also very often added to cured tobacco
leaves for flavoring and to serve as preservatives (see also chapter 4,
where we discuss adding sugar to food and drink). R. J. Reynolds perfected this formula in 1913 with its much-hyped Camel brand, which
combined flue-cured bright and air-cured burley tobaccos. The flue
cured was sugary, but additional sugars in the form of licorice, invert
sugar, and others were added to the burley, exploiting the more porous
structure of that leaf. Camel, with its sugary "American blend," took the
market by storm and quickly became the model for most subsequent
brands throughout the world, especially when merged with Madison
Avenue savvy.

Cigarettes were sold using many different tricks: colorful labels and
trading cards, of course, but also ads in the newly emerging popular
magazines, eventually glossy and in color, many of which would come
to depend on cigarette advertising revenues (and so tended to slight
any discussion of tobacco harms). Medical publications such as the
*Journal of the American Medical Association* started publishing ads for
tobacco products in the 1930s, accompanied by claims for one brand
or another being "kind to your throat" (Lucky Strike) or "just what
the doctor ordered" (L&M). Smoking Camels was supposed to "give
your throat a vacation," while Old Gold offered "not a cough in a car-
load," and Pall Mall would "guard against throat scratch." And Vice-
roy offered "double-barreled health protection," while "more doctors"
were said to smoke Camel.[29] Jingles were often key to a cigarette's success, first in magazines and newspapers, and later on radio (from the
'20s) and on television (from the late '40s). The American Tobacco
Company also played on growing fears of obesity by urging especially

female smokers to "switch to a Lucky instead of a sweet" (a campaign from 1927). Psychologists hired by the larger firms sought to realign popular tastes and even clothing fashions to match the colors of cigarette packaging—as when American Tobacco sponsored green fall fashion balls and luncheons, all in green themes, to create demand for its Lucky Strike brand.[30] Women had complained that the packaging clashed with contemporary fashions, which prompted the manufacturer not to change its pack but to redesign fashions.

## "The Perfect Type of a Perfect Pleasure"

The cigarette is the deadliest artifact in the history of human civilization. In the United States alone, cigarettes still kill about 480,000 Americans every year, more than 15 times the number that die from automobile accidents. Such a large scale of death could not be possible without mechanization and its "improvement" over the course of the twentieth century.

The cigarette is also, however, a quintessential packaged pleasure with origins in the late nineteenth century. Mechanization dramatically lowered prices while also creating an excess of supply that, when filtered through mass marketing, created addictive overconsumption. Removing nicotine from cigarettes would cure the addiction, but we cannot blame that molecule alone, or any of the industry's elaborate chemical machinations. Marketing has also been crucial for the cigarette's success, since the brands of different manufacturers have been essentially indistinguishable, apart from the colorful packs in which they have been housed and the slogans to which they have been attached. The cigarette itself has always been symbolically plastic: it is neither inherently masculine nor feminine nor sexy or sophisticated, but only so by virtue of the labors of Madison Avenue. There is nothing inherently cowboyish about a Marlboro or feminine about Virginia Slims. Marlboro was in fact a woman's cigarette until 1955, when Philip Morris introduced the hard pack "flip-top" box to make that brand stand up to cancer. (Prior to its "Marlboro Country" campaign, Philip Morris had marketed that brand as "Mild as May" with "ivory tips to protect the lips," and even used babies to appeal to mothers.) In modern Korea, Virginia Slims are smoked only by men, which only seems strange if we have been fooled into taking market propaganda for any kind of truth.

Physical and symbolic packaging—marketing—have long been key to fashioning the desire to buy cigarettes and, like other engineered pleasures, has its roots in the late nineteenth century. Historians have shown how "engineering consent" would become integral to selling cigarettes and how public relations firms helped reassure smokers whenever scientists would publish evidence of harms.[31] The tobacco industry's formerly secret documents, disgorged as a result of litigation, reveal the crucial importance of packaging as a vehicle for cigarette symbols—just as the twenty-cigarette pack houses a daily dose of nicotine. William Dunn, head of psychopharmacology at Philip Morris (a.k.a., "the Nicotine Kid"), in 1972 was explicit on this, asking his colleagues to think of the cigarette pack as "a storage container for a day's supply of nicotine" and the cigarette as "a dispenser for a dose unit of nicotine."[32]

Packaging is crucial for cigarettes in several different respects. One of these is the basic sameness of all cigarettes: all are essentially identical in consequence of mechanization, but also from the perceived need by companies to produce a minimally "satisfying" (that is, addictive) product. Cigarettes today are all about three inches long and contain about 2 percent nicotine by weight. Flue-cured tobacco is almost universally used (to facilitate inhalation), and humectants are added to prolong shelf life. Cigarettes also don't "taste" profoundly different, because the whole point is inhalation—and humans don't have taste buds in their lungs. Some compounds are added "to make the poison go down easier" (this would include menthol, cocoa, and myriad sugars) while others confer an attractive pack aroma. But most common brand distinctions are flimflam or preposterous or worse, as when "filters" or "lights" are advertised as safer—they aren't.[33]

Packaging has been used to create the illusion of differences in this basic homogeneity, as was true for many other name-brand goods that emerged at the end of the nineteenth century. Cigarette manufacturers talk about the various "layers" of a cigarette: the nicotine as the core or ultimate goal, the tobacco burned to transport that alkaloid, the paper

**Figure 3.4** Ad for Liggett brand Fatima cigarettes from 1913, combining three key themes of early cigarette advertising: luxury, romance, and Orientalism. Hints of youth and athleticism (skating) are also present. For several thousand more images of this sort including downloadable color images and videos, see http://tobacco.stanford.edu/tobacco_main/index.php.

sheath used to house the chopped leaf, the pack that contains and advertises some particular brand, the carton containing ten packs, the case enclosing some number of cartons, and so forth. Cigarette makers also talk about their evolution from being sellers of tobacco (in the nineteenth century) to sellers of cigarettes (in the twentieth) to sellers of smoke (in the 1950s) to sellers of nicotine (from the 1970s onward). R. J. Reynolds's director of corporate research in 1972 described the industry as follows:

> In a sense, the tobacco industry may be thought of as being a specialized, highly ritualized and stylized segment of the pharmaceutical industry. Tobacco products, uniquely, contain and deliver nicotine, a potent drug with a variety of physiological effects.[34]

These druglike effects cannot be sold without imagery, and specifically the layered imagery provided by packaging. As William Dunn pointed out (in a secret internal document):

> The cigarette should not be construed as a product but a package. The product is nicotine. Think of a puff of smoke as the vehicle for nicotine. . . . Think of the cigarette as the dispenser for a dose unit of nicotine. . . . Smoke is beyond question the most optimised vehicle of nicotine and the cigarette the most optimized dispenser of smoke.[35]

Of course there are aesthetic aspects of any such addiction: Oscar Wilde already in 1891 described the cigarette as "the perfect type of a perfect pleasure": "It is exquisite, and it leaves one unsatisfied. What more can one want?" British American Tobacco executives in 1984 quoted those same lines from Wilde, clearly appreciating the value of this alkaloid insatiability: "Let us provide the exquisiteness, and hope that they, our consumers, continue to remain unsatisfied. All we would want then is a larger bag to carry the money to the bank."[36]

Here is the rub: the modern cigarette, cranked out cheaply by the billions, made attractive through the dark arts of Madison Avenue, perfected through the agency of legions of chemists, is a perfect, albeit extreme, packaged pleasure. With their mild, inhalable, smoke, cigarettes did very much for tobacco what the syringe did for opium or soda did for sweetness—with the added irony that much of the effort that went into perfecting the cigarette involved a quest to make it ever more "mild." The American-blend cigarette, with its bright flue-cured tobacco, was supposed to overcome "irritation," with the paradox that by virtue of being mild, the modern cigarette was actually easier to inhale and therefore far more deadly—by a wide margin. Milder tobacco smoke also made it easier for young people to indulge and harder to quit, since inhaling made smokers profoundly more addicted. U.S. Surgeon General C. Everett Koop in 1988 recognized what the tobacco industry had known for many years: cigarettes are as addictive as heroin and cocaine.[37]

Smoking in this new form also moved the habit out of the orbit of the occasional indulgence—often attached to some social ritual—and into the realm of an individualized, compulsive addiction. Most tobacco smokers prior to the twentieth century were probably not addicted; their consumption was not sufficiently high, their indulgence not sufficiently regular. And they were not inhaling. Smokers would smoke ritualistically, to celebrate a wedding or business deal or birth of a child or a holiday; think also of the ceremonial Indian "peace pipe." Smokers of cigars would also not be smoking constantly throughout the day, and every day the same amount; there must have been such smokers but they were the exception, not the rule. Sales of cigars reflected this episodic, seasonal use: sales would rise dramatically in December, for example, as smokers prepared for the holidays.

Cigarette consumption, however, is quite different. There is nothing ritualistic or celebratory about their use. Cigarette smokers smoke roughly the same number of cigarettes every day, and the product is not used to celebrate anything, Hollywood fantasies notwithstanding. Rather, it is used compulsively to satisfy a physical craving as strong as that produced by refined opiates. Smokers of cigarettes have compromised their ability to choose; they are in the grip of the nicotine molecule, which essentially rewires the brain to create a powerful compulsion.[38]

Cigarette makers also realize that what they are selling is not so much a pleasurable experience as a set of images, associations, and illusions.

That is why packaging has become so important. For cigarette makers, packaging and specifically package art has steadily grown in significance, as manufacturers have lost access to most of their traditional advertising venues. Once commonly advertised on radio, television, and billboards, cigarette makers in most parts of the world now find it harder and harder to advertise outside the pack. In the United States, advertising on broadcast media was banned in 1970, followed by bans on ad implants in movies, vending machines, and even point of sale ads (including so-called power walls). The pack itself has become a kind of last refuge for advertisers — along with point of sale and direct marketing by mail — which is why so much effort has been put into combating laws mandating the sale of cigarettes only in plain drab packaging. Australia passed the world's first law requiring plain packaging in 2011, and cigarette makers have fought quite hard to have that law thrown out of court, using the World Trade Organization and other instrumentalities as a platform. The industry obviously sees this as a crucial battle, given that once pack adverts vanish, there will be little to differentiate brands and little to excite the imagination of young "replacement" smokers (a.k.a., "rookies," "novices," "pre-smokers," and "learners," in the industry's jargon). Other venues will no doubt be found to advertise (Facebook and other social media, for example), but the loss of the persuasive pack is clearly something cigarette makers don't want to see.

The cigarette is an extreme case of the packaged pleasure, and its history and psychopharmacology reveals one important qualification of "pleasures" that turn into addictions. The manufactured cigarette for most smokers today does not in fact create pleasure; it is rather a habit — a compulsion — that most wish they could leave behind. Smokers do not smoke because they like it, they smoke because they are addicted and cannot quit. Surveys repeatedly show that most smokers do not like the fact they smoke; indeed, smokers who say they enjoy smoking are so rare that the industry calls them "enjoyers" and "rather unique."[39] This makes smoking very different from, say, drinking. Most people who drink do so in moderation; they are not addicted. Only about 5 percent of people who drink alcohol are alcoholics, whereas most of those who smoke (cigarettes) are addicted. The difference is crucial: it means that whereas most people can enjoy alcohol as a recreational drug, smoking is very different. Tobacco is not a recreational drug. Smokers of cigarettes smoke because they feel they cannot do otherwise; they have no choice.[40] Robert Bexon, marketing VP and later president of Imperial Tobacco Ltd. of Canada, in 1984 put the matter as follows, after comparing smokers not to drinkers but to alcoholics:

> If our product was not addictive we would not sell a cigarette next week in spite of these positive psychological attributes. . . . The physical/sensory properties of smoking (taste/pleasure/like it) do not even surface in the smoker's description of why he smokes. It is actually seen as an unpleasant sensory experience. . . . Like alcoholics smokers realize they will always be smokers and can always fall off the wagon.

In contrast to the packages that will be discussed in subsequent chapters, "pleasure" in this sense is no longer even what is being packaged in cigarettes. A century and a half of mechanization, marketing, chemical manipulation, and corporate duplicity have turned most use into abuse, a luxury into an addiction, a pleasure into a deadly burden.[41]

# 4

# Superfoods and the Engineered Origins of the Modern Sweet Tooth

Food has always been about more than survival, more than about converting carbohydrates, fats, and proteins into flesh, bone, and energy. Foodways are like windows into a culture and tell us much about what a people values—and how knowledge and power are deployed or suppressed. There is so much choice in this realm—ingredients, ways of cooking (or not), mixtures, order and style of presentation, when and with whom to partake—that eating becomes one of the most expressive and ritualized forms of daily life.[1]

Foods can also stimulate the senses and will often include chemicals that modify human mental states and behavior. We have already commented on how psychoactive substances such as fermented alcohol and plant opiates were intensified in the form of distilled spirits and morphine, and how tobacco consumed in the form of cigarettes increased the potency and harm of older forms of nicotine use, but it is also important to realize that foods have undergone a similar process of intensification, as new technologies have enabled the manipulation of foods in ways not previously possible. Especially in the latter part of the nineteenth century, food engineers built on older discoveries of sugar, chocolate, and other flavor additives, along with carbonated water and diverse processed oils, to produce a wide range of sweets, "soft" drinks, and snack foods that were usually energy rich (if nutrition poor). Natural plant and animal products, especially sugars and fats, were cured and combined into potent pleasures transcending traditional foods. Often based on simple, if historically rare, ingredients, these newly concocted foods took an astonishingly wide range of forms. Sugar could be boiled, spun into cotton candy, "burnt" (becoming caramel), or combined into near-infinite forms, putting the humble (but healthful) apple to shame. Chocolate beans and kola nuts but also potatoes and corn and even animal byproducts were transformed into energy and flavor "hits," creating new physical delights as well as new sites of moral anxiety and new pathological conditions in the human body from overconsumption.

Sugar consumption was surprisingly rare prior to the packaged pleasure revolution, with American annual per capita intake (for added sugar) being only about six pounds (in 1822) compared with the hundred-odd pounds (!) achieved by the end of the twentieth century. What has changed are the techniques by which such "treats" are produced as well as the intensity with which they are marketed. This is important to keep in mind when thinking about whom we might blame for our modern epidemic of obesity. All too often we tend to blame the victim—the overweight are chastised for the twin sins of gluttony and sloth. But obesity is as much an environmental (and political) disease as anything else, and we therefore have to look at the kinds of foods made available and the forces pushing us to consume them. And for this we must understand not just personal sins but changes in how such foods are produced and marketed.[2]

Understanding the historicity of foods—and especially those loaded with sugar—means, among other things, paying attention to how technologies impact the quantity and quality of what we eat. The first crucial fact to appreciate, though, is the dramatic increase in variety of foods made available over the last couple hundred years. Until the 1860s, in fact, European peasants ate little apart from cereals such as wheat, oats, rye, and barley, along with root vegetables such as turnips, rutabagas, and potatoes, and pulses (peas, beans, and lentils), occasionally flavored with meat and cheese. Fresh vegetables and fruits were generally available only seasonally or, if preserved, in dried forms missing much of their nutritional value and original taste. Much of this food was also roughage and hard to digest, and made palatable only when spiced, seasoned or fermented.

Of course there have always been class differences in foodways, accentuated by the fact that for nonelites prior to the nineteenth century, food might cost up to 70 percent of a family's household income. The rich could always pay for variety: in medieval Europe the well-to-do had a taste for lean meats and fish doused in spicy, vinegary sauces, and sixteenth-century aristocrats took a fancy to foods rich in fat from oils and butter made famous by French chefs. Shipping of refined sugar and spices from distant lands also created new food fads and fashions. Exotic luxuries eventually trickled down the social scale, allowing eating to become a form of recreation or break from routine—even for the less well off. From the 1770s on, tea, coffee, and cocoa were at least occasionally affordable even for European commoners, albeit often in diluted, adulterated, or substitute forms, as when chicory replaced real coffee or sugar lovers had to settle for molasses. Spices, sugar, and other taste or energy boosters were major leaps. Spices and other new foods "assumed dreamlike qualities, those same qualities that characterized the distant and mysterious Orient: an 'oneiric horizon' onto which westerners projected their desires and utopias."[3]

Much of this increase in variety, however, is surprisingly recent. Potatoes were not a big crop in Europe until the eighteenth century, for example, which is also about when rice was introduced into the United States. Oat, millet, and buckwheat breads gave way to softer breads made from rye and wheat. For centuries, grains dominated the diets of ordinary Europeans, accounting for between half and three-quarters of total caloric intake as recently as 1800. Increasing wealth (and trade) eventually allowed higher-fat diets to reach the masses—which for a time at least was not a bad thing. Lack of fats (containing the essential vitamin D) explains why rickets and other crippling deformities persisted among the poor; and the substitution of butter and oil for water in cooking was at first beneficial. Indeed, the dramatic increase in access to butter, meat, and eventually margarine in the final decades of the nineteenth century began a dietary democracy that helped to reduce the gap in health and life expectancy between rich and poor.[4]

But underlying these advances were other changes. A new category of comestibles was beginning to emerge, especially in the nineteenth century—we shall be calling them *superfoods*—based mostly on sugar and other exotic plants processed and often intensified by industrial means. *Hyperfoods* might be a better expression insofar as such products caused a "sugar rush"—and were "hyped" in advertising to displace other foods. *Pseudofoods* is also apt, given that most lacked the protein, vitamins, roughage, and minerals of meats and vegetables. And even if not inebriating or psychoactive in the manner of older intoxicants, they did offer hits of pleasurable taste and surges of energy at increasingly affordable prices. As packaged pleasures they were taken into the body in shots of intensity, quintessentially in the form of flavored sugars juiced up with carbonized water, ice, or fat (as in ice cream). Foods of this novel sort quickly flourished in an extraordinary variety of forms, adding texture, color, and flavor to the otherwise bland diets of common folk—with the crucial innovations coming precisely in this era of the packaged pleasure revolution.

Superfoods stimulated, of course, but over time they would also come to crowd out more substantial nourishments, displacing vitamins, minerals, proteins, or roughage in the popular diet. Often they would become supplements to regular meals, "pick-me-ups" or "snacks" relieving the routine and fatigue of the daily grind. And in time they moved from being rarities to daily indulgences, consumed to relieve the tedium of home or office work or industrial labor. Superfoods also attracted those who were traditionally at the foot of the dinner table—women and children. Limited in their access to the potent tastes of alcohol, tobacco, and even roasted meat, women often embraced these foods as guilty pleasures. And on the street away from the chastising eyes of parents and family, children could gain a kind of freedom in the packaged pleasures of candy, soda, and ice cream—albeit for a price.

## Sugar

Superfoods begin with the allure of sugar. Despite its natural rarity, few human cultures have ever rejected the introduction of sugar into their traditional diets. Most primates crave sugar because sweetness generally indicates edibility. As fruit-loving primates, humans have evolved to enjoy the sweetest and therefore the ripest fruits—which often also have a high vitamin content. (No sweet fruits are poisonous, interestingly.) In caloric terms, all sugars are quite similar—simple carbohydrates that are easily absorbed and converted into energy—and even complex carbohydrates like starches are broken down into sugars in the digestive tract. Sweetness is the first of the four taste sensations in the mouth to reach the brain (the others being salt, sour, and bitter). Babies prefer sucrose and fructose to less sweet lactose and glucose, and the craving for sweetness fades somewhat with age as taste buds mature.[5]

Sugar love is more than a biologically driven craving, however. It is among the easiest goods to transport and keeps almost indefinitely; it is also a preservative, and plays a crucial role in the fermentation of beer and wine. It can also be used to impede crystallization when added to frozen foods. There are a lot of reasons to favor sugar, even if it leads to tooth decay, weight gain, diabetes, fatty liver, and a host of other maladies.[6]

Yet sugar has come only recently into the human diet as anything more than a rare and exotic treat. Sugar is probably the best modern example of a pleasure based on natural scarcity, and one that clearly endangers us when it becomes too readily available and too heavily promoted. (There is a powerful sugar lobby in the United States combating taxes on soft drinks and protecting subsidies.[7]) Of course, there are precedents in nature: humans have presumably always cherished the sweetness of honey, for example, a trait we share with badgers, bears, and a number of other mammals. Honey is beloved in most historical traditions: for ancient Mediterraneans honey was the food of the gods; and in Vedic and Scandinavian legends honey was considered a heavenly libation and a charm against evil. Babylonians consecrated the ground by sprinkling it with honey and wine, and honey has even been used to preserve the corporeal remains of honored dead. Alexander the Great was encased in honey for his long journey back to Greece, following his death in Babylon in 323 BCE. Ancient Chinese sweet and sour dishes used maltose from grains and sorghum grass as sweeteners. Every people on earth has valued sweets, whether in the form of the sweet-tasting "honey ants" of aboriginal Australia or the palm and coconut sugars savored in Southeast Asia.[8]

Sugars are found most prominently in the juice of green plants. Only in sugarcane and sugar beets, however, was there enough to be harvested and refined on a large scale commercially — until the explosive development of corn sweeteners in the twentieth century. Cane sugar has an older history: the people of New Guinea chewed wild sugarcane as long ago as 8000 BCE, and the plant seems to have been brought to the Philippines and India about 6000 BCE. But it was only around 500 BCE that people on the Indian subcontinent developed methods by which a crude form of crystalline sucrose could be made by evaporating juices pressed from sugarcane. Cane reached China by the third century BCE, though it was not cultivated there until much later. Persians saw similar delays: soldiers from the invading armies of Emperor Darius found "the reed which gives honey without bees" growing along the Indus River of India in 510 BCE, but it was not until the sixth century CE that Persians began processing cane into sugar. Arabs after entering Persia in 642 CE returned with sugar and its refining secrets — and by 755 CE were cultivating sugarcane in North Africa and Spain.

Still, the spread of cane was limited by how difficult it was to cultivate. Sugarcane has a long growing period, often upward of twelve months prior to harvest, requiring much fresh water and a tropical climate. And though Venice had imported Muslim sugar from 966 CE, Europeans were mostly content to purchase refined sugar from the Middle East until the Venetians in 1470 started importing cane and its refining techniques, expanding markets to the north. Though rare on even the tables of aristocrats, some medieval recipes reveal it being used to make hard candy. And luxurious banquets featured candied violets, roses and even marigolds, made by soaking sugar solutions into the petals. Sugar was also taken as a medicine, as was the habit with spices of a rarer sort. Humoral theories classified sugar as hot and moist, and therefore therapeutic in persons thought to be suffering from an excess of humors cold and dry.[9]

The "spice" appeal of crystalline sugar is understandable, given that it can easily be mixed with fruits, coffee, chocolate, cinnamon, and vanilla without compromising the native taste of the sweetened product. In fact, as when mixed with cocoa, sugar dramatically enhances taste by removing bitterness. Honey, by contrast, has a distinct savor and, dependent as it is on the diligence of bees, cannot be easily mass produced and thus could never become the foundation of a broad-based sweetening revolution.[10]

But sugar was not at first easy to manufacture. Within a day of cutting, the cane had to be crushed and squeezed to yield sweet juice; the liquid then had to be boiled to crystallize. Some skill was required to know precisely when to "strike," that is, to remove the boiling sugar from the heat at just the right temperature for crystallization to occur. The resulting raw sugar was a gravelly brown (our English term actually derives from a Sanskrit word—*shakara*—meaning "gravel") and usually still harbored contaminants in the form of dirt, mold, bacteria, or even lice. Refined sugar results from repeated stages of boiling and drying, using lime (or egg white) to coagulate and float off impurities that can then be skimmed away. A middle stage in this refining process was blackstrap molasses, with its own myriad uses. After the purified juice was boiled and strained, the resulting syrup was conveyed into earthen molds and dried. The residue was then mixed with water and milk and boiled and dried once again to make white sugar in pure crystalline form. This cumbersome process required much wood fuel and fresh water—not always abundant in cane-growing regions. Which is one reason raw sugar was shipped to Venice and Antwerp for refining in the fifteenth century.[11]

For centuries, high prices encouraged the spread of sugarcane cultivation far from its ultimate consumption, especially in Europe but also later the Americas. In the 1390s, sugar plantations expanded into southern Spain and by 1420 had reached the Canary Islands, Madeira, and the Azores off the Iberian coast. Cane then followed Columbus into the Americas, appearing in the Caribbean in 1494, Mesoamerica in 1520, and Brazil in 1526. The English could not grow sugarcane in

**Figure 4.1** An English lithograph by John Hinton (1749) of a West Indies sugar plantation and slaves crushing the cane and boiling the resulting juice. Library of Congress (LC–USZ62–7841).

Jamestown — and not for lack of trying — but they did create a thriving trade in sugar (and rum) from the Caribbean island of Barbados by 1627. The French colony known as Saint-Domingue (subsequently the independent nation of Haiti) became the largest sugar producer in the world in the eighteenth century, based almost entirely on slave labor. Planters were obliged to exploit new lands like Cuba and colonies in Africa and islands in the Pacific, notably Fiji, once it was realized how quickly cane could exhaust the soil.

The sugar trade yielded high profits well into the eighteenth century, so much so that the sweetener was called "white gold" and was heavily taxed. But industrialization and mechanization eventually lowered prices. In 1768, sugar milling was mechanized with the invention of steam-driven crushers, and in 1813 Edward Howard of England developed a closed steam-heated vessel which, along with other innovations, reduced the time required for refining from two weeks to only twenty-four hours.[12]

As a tropical plant, European access to sugarcane was always vulnerable to naval blockades, as the French learned painfully during the Napoleonic wars. Though the German Andreas Maggraf identified the possibility of extracting sucrose from a beet root as early as 1747, Europeans developed this alternative source only when Napoleon banned cane sugar from those parts of Europe he controlled in 1813. Today beets provide about 30 percent of the world's total sugar production.[13]

Another source of sweetener came from America's massive bounty of corn, facilitated by the growth of rail transport and the invention of grain elevators. In 1856, when the Chicago Board of Trade began to grade corn (i.e., maize) for purity and water content, corn became a virtual river that could be channeled into different conduits of processing. Corn oil became a significant (and mobile) commodity,[14] but corn refiners soon also discovered (in 1866) how to transform corn starch (from the crushed and dried endosperm) into corn syrup, a convenient sugar substitute.[15] In the 1950s, biochemical innovators learned how to transform glucose-based corn syrup into high-fructose corn syrup, a process that, when industrialized in the 1970s, yielded a sweetener much cheaper than traditional sucrose cane or beet sugar. The net effect was a massive increase in the consumption of high-fructose corn syrup, in sodas of course but in many other foods as well—one of the principal causes of our ongoing obesity epidemic.[16]

But how exactly did sugar become a popular packaged pleasure?

## Packaged Sweets

Once an exclusive indulgence of the rich, sugar had found its way onto the tables of the poor of England and the United States by 1800. Teaspoon after teaspoon was poured into tea and coffee cups and spread on bread as jam. Sidney Mintz notes that the turn to sugar (and, we should add, fats) most often has meant a move away from diets based on a single complex carbohydrate—such as rice, wheat, corn, potatoes, or millet served with a "flavor-fringe supplement," as in chili pepper or a bit of fish.[17] Beverages also shifted toward sugar-based drinks—sweetened rum of course but also highly sugared tea and coffee. By 1856, refined sugar consumption in England was forty times what it had been a century and a half earlier. In the 1840s alone, improvements in sugar refining halved the price of white sugar.[18] Cheap sugar made it possible to elevate the caloric intake of the poor without increasing their consumption of meat, fish, poultry, or dairy, and sugar became a kind of entitlement. There was always room for dessert, and by the Second World War American soldiers would be offered chocolate bars (as well as Coca-Cola and cigarettes) as compensation for their sacrifice.[19]

While sugar has been put to many purposes, its quintessential use has been in candy. Sugar is everything in candy. Nineteenth century candy-making manuals popularized the basics. Boiled to 295–310°F (the brittle "hard crack" stage) sugar will produce hard candy, higher temperatures (the "brown liquid" stage) are necessary for caramels, and lower temperatures will yield "fondant" candy, a creamy white substance used in soft candies. Many candies (notably kisses and taffy) require temperatures somewhere between 240° and 295°. And pulling, spinning, stirring, or otherwise working cooked sugar will produce distinct textures. "Doctors" like cream of tartar, corn syrup, and fruit acids are added to reduce crystallization or to otherwise improve tastes and textures.[20]

Different forms of sugar—brown, refined white, powdered—produce different textures and possibilities for mixture with a near-infinite variety of flavorants and colorings. Flavors come from "essential oils" distilled or squeezed from fruit peels, fragrant bark such as cinnamon, and spicy seeds such as aniseed, caraway, or coriander. Citric acid from fruits makes for a tart taste, but this is only the beginning. To basic hard candy, dried fruit or nuts can be added. Hard candy can be pressed into starch molds of any shape or formed into tablets or lozenges in hollow metal molds. The all-day sucker is just a tube of colored hard candy sliced in cross-section into a series of narrow discs and put on a stick. Adding cream and nuts to boiled sugar (along with corn syrup and brown sugar) makes English toffee. To basic fondant, the addition of butter, bitter chocolate, and nuts makes fudge. Albumen, made of egg whites, combined with aerated sugar becomes nougat or, if cooked at a higher temperature, marshmallows. Pectin, a fruit polysaccharide isolated first in 1825, is a gelling agent added to sugar to form jellies and jams, while vegetable starch makes possible jelly beans and candy corn. Sugar is a remarkably malleable substance, and many of these techniques were discovered in the nineteenth century.[21]

Crucial to appreciate, though, is how radically mechanization transformed candy making and candy consumption. Given sugar's early status as a medicine, innovation often occurred in pharmacies. In 1847, a Boston druggist by the name of Oliver Chase developed a machine consisting of two rollers into which pill-shaped molds had been cut, allowing the production of bite-sized candy "tablets" when sheets of boiled sugar, gum arabic, and flavoring were fed into the rollers. Chase's efforts eventually led to the founding of the Necco Candy Company, whose rolls of flavored candy wafers would attract many generations of American kids. Chocolate coating, or "enrobing," machines (as they were called, magisterially) appeared around 1870. Hard candy making was aided with an Austrian invention, a continuous vacuum cooking apparatus, in 1906; a mechanical candy cutter debuted in 1908, and a machine for filling hard candy shells with soft centers appeared in 1910.[22]

Novel textures were added to this mix, as with the improbable invention of chewing gum—a confection that offered a bit of flavored sugar in the form of a chewable yet indigestible gob. Here was an opportunity to masticate without having to buy actual food, appealing to those convinced of the value of exercising the muscles of the jaw—which was also held to aid in digestion or the relief of boredom or tension. Chewing gum was advertised as a convenient pleasure, delivering a dose of flavored sugar without the bother or bulk of an apple or pear. It was principally about taste, of course, usually mint, pepsin, or more mysterious flavors like "juicy fruit" or whatever was in bubble gum.[23]

Modified from a habit of American aboriginal peoples, chewing gum was first made commercially in Maine around 1850 from spruce tree resin—and later from paraffin wax. Chicle became the chewable base around 1890, after tests found the substance (first used as a rubber substitute) more chewable and better able to carry flavoring than wax or spruce resin. Making the stuff was easy, and competitors offered a wide range of flavors. As with so many packaged pleasures, however, advertising and economies of scale quickly narrowed the field of producers. The biggest turned out to be William Wrigley, a Philadelphia manufacturer who, after moving to Chicago, built a business selling baking powder. When offering chewing gum as a premium gift in 1891, he found his new promotion more popular than his baking powder and shifted over to gum making, introducing Juicy Fruit and Spearmint in 1893. Business boomed when he launched a massive advertising campaign in 1906. In an effort to exploit the link with daintier sweets like bonbons and breath mints, Wrigley appealed to women in ads that promoted his gum as an after-dinner pleasure that should always be on hand and purchased by the box. Ads in family and women's magazines promised a "calming effect" and guaranteed "white teeth and sweet breath." Beeman's Pepsin gum went even further when it claimed to cure seasickness. Another company insisted that a stick could digest four ounces of meat, and mothers were encouraged to serve gum after every meal to aid digestion.[24]

Wrigley quickly recognized that the biggest potential market would be with footloose youth and children, and so he began to advertise on billboards and in electric lights on Times Square. The company went so far as to mail two sticks to two-year-olds on their birthdays. Besides trying to create new consumers among the young, Wrigley insisted on making his five-stick packages available everywhere for chewing at all times. In stores and bars he insisted that Juicy Fruit be sold next to the cash register to attract the spare nickels of impulse consumers, some of whom may have hoped to hide their alcohol breath from disapproving wives and mothers.[25]

Despite the seeming crudity of chewing gum, the habit won wide acceptance in early twentieth-century America. Michael Redclift notes that immigrants embraced the chewy stuff as a cheap way of assimilating to the American way of life. More to the point, chewing gum directly and emphatically served the purpose of many other sugared treats and snacks, offering quick energy and flavor without even pretending to be a food—a quintessential expression of the sugar revolution.[26]

Another curious treat was equally a product of American ingenuity

**Figure 4.2** Still touting chewing gum as a digestive aid in the 1920s in full page ads, Wrigley's playfully used its "arrow" theme in its trademark to promote brand identity. *Saturday Evening Post*, May 12, 1928, 108.

and culinary "daring" — the irrepressible Jell-O. Jell-O was an early example of a manufactured pseudofood with no previous history in the American (or human) diet (other examples include Cool Whip, Spam, Velveeta, and Cheez Whiz). Even within a food industry marked by extensive processing, Jell-O was an amazing product. Based on collagen derived from slaughterhouse waste (bones, connective tissue, intestines, and skin), gelatin is a purely engineered (if improbable) pleasure. With added sugar, like so many things, it became a treat.

The gelatin underlying Jell-O was invented in 1845 by Peter Cooper, the famed manufacturer of locomotives, promoter of an Atlantic telegraph cable, and founder (in 1859) of the Cooper Union, a (once) tuition-free college open to both sexes in lower Manhattan. This jiggly substance was improved by the Knox Company of Johnstown, New York, in 1893 and then again by Pearle Wait, a maker of patent medicines, who added flavor. In 1896, Wait's fruity formula was bought by the Genesee Pure Food Company owned by Orator Woodward, famous as the maker of Grain-O, a health drink without caffeine. The Genesee Company joined other pioneers in the world of processed foods by creating a mass market for a product no one had ever imagined they might want or need. Woodward's packaged flavored gelatin was sold in powder form as Jell-O, with advertising directed to women to appreciate this new "food" as a convenient, "dainty," and healthy (because it was nonfat) alternative to fruit pies, custards, and other rich and hard-to-make desserts. Genesee offered homemakers an amazing variety of uses for Jell-O, including mixing it with shredded wheat in a sandwich — one of many suggestions that never caught on.[27]

Jell-O was immediately and immensely popular, however, with some twelve million recipe booklets featuring the slippery stuff distributed by 1915. Children and their parents were a principal market target, with ads remarking on how Jell-O had the flavor, consistency, and sweetness cherished by kids. Jell-O's ads often used images of delighted and delightful children, including the blonde, cherubic daughter of the ad agency's head, riding a new wave of child-centered consumption. Ads promised parents that Jell-O could overcome the resistance of kids with "fastidious appetites" at a time when parents were beginning to tolerate fussy children. Genesee Pure Food warned anxious parents to avoid the "dreadful disappointment" that "you remember as a child" by serving what the kids really want: Jell-O. And when Campbell's rolled out its spunky Kids (to advertise soup), Jell-O bought the rights to use Rose O'Neill's Kewpies (cutsey angelic figures) to promote their sweetened gelatin. Opera star Ernestine Schumann-Heink and stage and film actress Ethel Barrymore were also hired to hype Jell-O to aspiring middle-class homemakers. And like other mass-produced American inventions, Jell-O became a symbol of assimilation to millions of immigrants. Playing the patriotism card, Jell-O in the 1920s courted recent arrivals with a Norman Rockwell–illustrated cookbook in different languages.[28]

Of course, Americans didn't buy Jell-O just because it was marketed cleverly and promoted as the "American way." Jell-O had basic sensual attractions: in its pure form it shimmered and could even be seen as playful as it jiggled and slithered down one's throat. Essentially a plastic imitation of food, it could also be molded into hundreds of toylike shapes while being adaptable to many different culinary inventions — by incorporating fruit or sliced vegetables, for example. And its proto-modernist gelatinous clarity combined well with the new (and often garish) synthetic food dyes put out by German chemical manufacturers. But like many other newly packaged foods, from a purely nutritive point of view Jell-O offered little more than sugar (and a bit of "food coloring") to the diet. And today, while fewer grandmothers may be offering Jell-O as a dessert for Sunday dinner, it remains popular with small children and their parents and has even sprouted into new food niches, as when frisky adults down Jell-O shots laced with liquor.

We should also keep in mind that for many years candy, unlike fermented beverages and tobacco, was associated with the cravings of women and children, with the added tint (or taint) of the frivolous or lavishly extravagant. White sugar (think frosting) was the main ingredient in the Victorian wedding cake, the centerpiece of the quintessential female-dominated festival. And while it is easy to forget that such things have origins, the remarkable fact is that it was not until the nineteenth century that sweets became an essential part of the celebratory rituals of Christmas, birthdays, Valentine's Day, Halloween, Easter, and (after 1914) Mother's Day — all holidays built around women and/or children

## The Kewpies' First Banquet

For a long time the Kewpies have been distributing sunshine and cheer and good times, but nobody has ever done anything for them beyond saying, "Aren't they cute?" Now the Jell-O Girl is giving them a banquet and is serving their favorite dish of

Every sensible woman will agree with the Kewpies that Jell-O is the proper thing to serve for dessert—not only because its flavor is delicious, but because it is so easily made up into the most delightful dishes without cooking and without adding anything but boiling water to the powder from the wonderful ten-cent package.

There are seven different pure fruit flavors of Jell-O: Strawberry, Raspberry, Lemon, Orange, Cherry, Peach, and Chocolate. Each, in a package by itself, **10** cents at any grocer's or any general store.

**A beautiful new Jell-O Book telling of a young bride's housekeeping experiences has just been issued. It has splendid pictures in colors and will interest every woman. It will be sent to you free if you will send us your name and address.**

THE GENESEE PURE FOOD CO., Le Roy, N. Y., and Bridgeburg, Ont.

A tightly sealed waxed paper bag, proof against moisture and air, encloses the Jell-O in each package.

This is the package

**Figure 4.3** Associations with the ubiquitous image of the Kewpie as well as the appeal of the easy-to-prepare dessert were used to sell Jell-O. *Saint Nicholas Magazine*, February 1916, 14.

and the gifting of sweets. Sweets also became central in the customs and rhetoric of courtship: from the simple "conversation lozenges" with their romantic messages printed in food coloring (appearing first in the 1860s) to the soft-centered chocolates and bonbons to be savored slowly, candy became part of the lover's vocabulary. While today we may take for granted the marketing of candies to children, this is relatively recent: as late as 1900 the more common pattern for candy was still to market more to women, for gifting and for afternoon teas.[29]

From the 1830s on, with the dramatic cheapening of sugar, confectioners began selling sweets to the less well off and then (eventually) to children who could now clamor for hard-boiled sweets for pleasure alone, absent any of the earlier intimations of sugar being medicinal. Candy, after all, provided "energy," a nineteenth century construct and obsession — and there was not yet any notion of "empty calories."[30]

Candy fads of course have multiple origins, but one main line stems from medicinal sweets and the sugarcoating of otherwise foul-tasting lozenges. What once cured pain could now serve unadulterated pleasure, and the healing "treatment" of the candy tablet was shortened to a "treat" for kids — an interesting etymology. Sold in pharmacies from large glass jars on the counter, these "penny candies" were placed at the eye level of six-year-olds, who would pester their parents or plunk down their own coins to acquire a handful. Jars of colorful candy created desire, which was probably only intensified when parents refused to give in. Confectioners offered imagination and novelty in their treats; penny candies were at first often called "toys" and tied to innocent juvenile pleasure. Some candies were even meant to be played with before being eaten — like the candy marbles made popular in the 1890s. When colored and shaped to look like gems, hard candies also toyed with children's fantasies, and some merchants gave away small toys or a miniature doll with each purchase, adding yet another dimension to the packaged pleasure.[31]

All this came long before the trinket was added to the Cracker Jack box in 1912. Even without a bonus, however, the appeal was more to novelty (or fantasy) than to quality. Lollypops, for example, were made into the form of rakes and hoes in the spring, and then into the shape of firecrackers for the Fourth of July. Kids seemed to like eating Seashore Pebbles that looked like rocks but turned soft in the mouth. Sun-Ray Bottles were cones of sugar filled with flavored syrup, and then there were the endless incarnations of licorice—as shoelaces, oddly enough, but also in the form of candy cigarettes, cigars, and pipes. Variety was the name of the game. A late-nineteenth-century store in Wheeling, West Virginia, boasted candy drops, cocoa nuts, candied almond creams, marshmallow drops, mint drops, cream bonbons, cinnamon drops, licorice drops, jelly gum, and chewing gum, all of which were new to the world in recent decades. After 1900, many of these same basic candies were repackaged and rebranded by companies, like Robert Welch's Sugar Daddy (1925), Hirschfield's Toostie Roll (1896), Henry Heide's Red Hot Dollars (a cinnamon hard candy from the 1920s) or Clarence Crane's simple boiled sweet with a hole in it, to which he gave the auspicious name "Lifesavers" in 1912.[32]

As harmless as these may have appeared to contemporary devotees, sweets also had a complement of worriers cautioning about the dangers in their overuse. Moralists warned that a young woman might be lured into an unsavory romance with a suitor who would falsely win her heart with a box of seductive sweets. Others feared that candy would undermine character development. As early as 1834 an American writer fretted in *The Friend* that the street child who indulged in candy was bound to lead a life of "intemperance, gluttony, and debauchery." In 1856, Amos Bronson Alcott, a New England educator and transcendentalist, advocated that children stick to fruits and vegetables, reasoning that overindulgence would lead to a life of drink (rum, after all, was made from the juice of sugarcane). James Redfield in 1856 worried that the habitual candy user "requires larger and still larger potations of sugar to satisfy him," going "from one candy-shop to another as the toper goes from one coffeehouse to another to satisfy himself with drams." Others feared that candy might lead children to inappropriately imitate adults—via chocolate cigarettes or candies made into the shape of liquor bottles labeled "gin," for example. A related danger was that candy and tobacco were very often sold in the same stores and in similar penny packs, which would often include cards depicting sports figures and other celebrities as added attractions.[33]

A lesser—albeit far more common—complaint was that sugar was starting to spoil the legitimate dinnertime appetite of children who should be eating "wholesome" foods. Complaints of this sort are reflected in the rising popularity of novel expressions in the English vernacular, with broad talk for the first time during this era of "spoiling your appetite" (1850s), "wholesome foods" (1880s), "dangerous foods" (1890s), and "balanced meals" (1910s).[34] Unrestrained candy eating was sometimes also seen as threatening the centrality of the family meal in daily life, a corrupting influence on both body and mind.

Complaints of this sort, however, did little to slow the candy train. Sugar paired well with the caffeine in tea and coffee and came to be widely used in jams and jellies. Sugar transformed morning breakfast (when sprinkled on cereals) and most forms of rituals where gifts were exchanged. It also became a key factor in most holidays and celebrations—all of which was new in this era of the packaged pleasure. Indeed, by the end of the nineteenth century it is hard to name a food or even tobacco product not sugared up to one degree or another.[35]

## The Allure of Chocolate

Like many other super- or hyperfoods, chocolate is mildly stimulating and perhaps even addictive to a certain extent. Cacao contains a small amount of caffeine (although not as much as coffee) but also theobromine, a bitter psychoactive alkaloid that, with caffeine, stimulates the central nervous system and dilates blood vessels.[36] Chocolate comes from an improbable source: a large green seed pod formed from a flower that sprouts directly from the trunk of a tree (*Theobroma cacao*) that originates in the American tropics. An insect called a midge need fertilize only four or five of a hundred flowers to produce the large oval pods, inside of which grow forty beans in a sticky white pulp. When mature, farmers scoop out the beans, which they ferment until dried. These cacao (or cocoa) beans are then roasted and crushed into a powder, which can be mixed with water and served as a cold unsweetened beverage. Around 2500 BCE the cocoa bean was brought from upper Amazonia to Mexico (though details of that route remain obscure), where it would eventually be enjoyed as the favored drink of aristocrats and elite warriors in the Aztec Empire. Like so many other superfood plants, its cultivation has spread far from its origins. Today most of the world's cocoa beans are grown in Africa, with only 1.5 percent being harvested in Mexico.[37]

Hernán Cortés introduced cocoa into Spain in 1525, believing it to be a "divine drink" — apparently because it relieved fatigue. But winning the European consumer was not so easy. Girolamo Benzoni in his 1575 *History of the New World* claimed that cocoa "seemed more a drink for pigs than a drink for humanity." In 1569, the Catholic Church allowed cocoa to be served during Lent, apparently because it was foul tasting and thus didn't seem to violate the sacrifice required of the faithful. Brought to France in 1615, it finally became the rage in the Paris court, and the drink reached England in 1655. The Spanish substituted cinnamon, pepper, and vanilla for the chili that Mexican Indians had been mixing into cocoa — and sugar was sometimes even added. Coveted as a luxury, it was often kept in pots of precious metal and porcelain. Cocoa was also prescribed as a medicine, stimulating cold and moist "humors" to counter the ill effects of "hot" fever. In the 1770s, chocolate was eaten as a paste of cocoa beans with added sugar and spices, including the rare and precious ambergris from whales. By 1789, Joseph Storrs Fry of England was using James Watt's steam engine to grind cacao beans; and even earlier, in 1765, the American James Baker of Massachusetts had successfully manufactured caked chocolate blocks. Still, despite these early lurches toward mass production, chocolate was generally identified with the idle clergy and nobility (though German poets also had a penchant). Even as a beverage, cocoa was drunk less as a routine morning stimulant and more as a luxury and delicacy of the "refined" palate.[38]

By contrast, tea and coffee became the daily beverages of the rising bourgeoisie — and tea had even reached the laboring masses in England by the 1790s. A century earlier London had boasted chocolate- as well as coffeehouses where politicos and gamblers whiled away the day. But coffee was the preferred temperance drink of the merchant and manufacturer. Originally from Ethiopia and other regions of eastern Africa, coffee beans were imported to southwestern Arabia in the thirteenth century and had become part of Yemeni Sufi culture by the late fifteenth century. It had arrived in Europe by 1517, mostly via the usual suspects (Venetian merchants) but also via war with Turkish armies.

As had happened with sugar, the eastern trade in coffee was eventually replaced by European colonial plantations. By the end of the seventeenth century coffee was being grown in Java for the Dutch, in the Antilles for the French, and in Central and South America for the Spanish and Portuguese. And like many other tropical botanical products, coffee was praised as having medicinal virtues prior to its broader incorporation into pleasure culture. London's first coffeehouse appeared in the mid-seventeenth century, and by the 1660s artisans, shopkeepers, and other achievers were beginning to substitute caffeine for beer or wine in the morning. Seventeenth-century coffee was served with milk and much sugar, providing energy as well as caffeine. The beverage came to be associated with the sober (yet alert) men of the counting houses about this time,[39] perhaps the first recorded imputation of the drink's mathematical virtues. (Alfréd Rényi, the great Hungarian mathematician, would later define mathematics as "a machine for turning coffee into theorems."[40])

Cocoa, though, was very different. Identified at first with the European aristocracy and American aborigines, the drink made from this bean for a time was seen by the rising European middle classes as simultaneously decadent and uncivilized. But cocoa was rescued from these contrasting censures by technological changes. Beyond its cultural associations, the biggest problem with the drink was its unpleasant oily taste from the vegetable fat in cocoa beans. The solution came with Conrad van Houten's invention of the hydraulic cocoa press (in 1828) that separated roasted beans (or nibs) into cocoa butter and defatted cocoa powder. This was a much more effective way to remove the oily butter than boiling and skimming it from the nibs. The fine powder could then be mixed in water to make a defatted chocolate drink (improved by adding alkaline salts in a process known as "Dutching"). A turning point of sorts came in 1866, when John Cadbury in Birmingham obtained Van Houten's machine and started marketing Cadbury's Cocoa Essence, a soluble powder for family use. Defatted cocoa, once liberated from the bitterness and cloying oil of the bean and sweetened, had a radically new taste. Chocolate drink, a mild stimulant, was thus born as a breakfast and children's beverage.[41]

Even more important, though, was the invention of the chocolate bar in 1847. Rodolphe Lindt in Switzerland and J. S. Fry in England blended cocoa butter and cocoa powder (from Van Houten's press) with sugar and molded the results into a bar. Milk chocolate was then invented in 1875 by the Swiss Daniel Peter, who added the condensed milk of Henri Nestlé to the mix. The smooth texture and mellow flavor we associate with "milk chocolate" was the result of Lindt's "conching machine" (1879), an ingenious device consisting of a shell-shaped granite bed over which rollers moved back and forth, grinding a mixture of cocoa powder, cocoa butter, sugar, and milk into a smooth-tasting paste. Conching eliminated the coarse and gritty "mouth feel" of the old chocolate. And chocolate consumption in the United States skyrocketed: from just over a million pounds in 1860 to about twenty-six million pounds by 1898.[42]

Sweetened, smoothed, mellowed, and wrapped, chocolate became the queen of packaged superfoods. Chocolate bonbons were boxed and had become associated with the rites of romance by the 1860s. High quality chocolates, offered in delicate (and artistic) parcels, became the "perfect" gift from Victorian beaus to their sweethearts. Part of the gender code of the time was the pairing of women with sweets (including sugared beverages, as we shall see) and men with "strong" drink (alcohol) and bitter stimulants, such as coffee. Chocolate's erstwhile link to aristocratic decadence was less a barrier for women than for the striving bourgeois male, just as extravagance in fashion (think lace and powdered wigs) survived in women but not in men (think of the Victorian businessman's sober, somber, and snail-pace changing vestments). Chocolate was a bit too sissy for real men—though it certainly could titillate. By the 1880s, links between the "primitive" sexuality of the tropics and chocolate had become an acceptable excitation on confectioners' trade cards, where bare-breasted female "natives" were featured gathering cocoa pods. Men of course could give chocolates, but women weren't really supposed to buy them for themselves—though of course they did. Chocolate became part of that ornate world of feminine refinement and romance, but also of guilty pleasure with vague

hints of naughtiness. Manufacturers routinely manipulated or sometimes even helped to create such images, and by the end of the century
had succeeded in creating an alternative association, linking chocolate
to virtuous and family-based manufactories: the Cadbury and Rown-
tree families in England with their Quaker and philanthropic links,
and Milton Hershey with his central Pennsylvanian Mennonite roots
and (from 1909) famous orphanage. But the naughtiness never disappeared entirely, and that was part of chocolate's deliberately crafted
attraction.[43]

Gradually, too, chocolate was passed on to children. By the end of
the 1800s, lighter-colored and sweeter chocolate was being made to appeal to the juvenile palate. Hershey advertised his milk chocolate bars
as helping to fashion strong and healthy young bodies, because they
contained milk, but other chords were struck as the stuff was molded
into chocolate cigarettes and cigars (Hershey sold "Hero of Manila
Chocolate Cigars," referencing Admiral Dewey's much-publicized role
in the Spanish American War of 1898). Other companies made milk
chocolate party favors in the shape of garden hoes, sabers, locomotives,
zoo animals, and, of course, Easter eggs and bunnies. Hershey then returned to the romantic theme with the Hershey "Kiss" in 1907.[44]

Milton Hershey (1857–1945) was the key figure responsible for popularizing milk chocolate in the United States. One of many independent
candy makers from central Pennsylvania, Hershey had started out stirring up batches of caramels in a small shop in Lancaster, Pennsylvania.
In 1893, he became mesmerized by a chocolate enrobing machine he
had seen at the Chicago World's Fair; he bought the machine from the
J. F. Lehmann Co. of Dresden to coat his caramels in chocolate. Soon
thereafter Hershey sold his caramel business in order to buy a dairy
farm near Harrisburg, Pennsylvania, to supply a new product, the milk
chocolate bar, in 1900. The area around Hershey's factory quickly grew
into a carefully planned company town, rechristened Hershey (from
Derry Church) in 1906. By 1915, Mr. Hershey owned sugar plantations
in Cuba to supply his candy making and had developed close ties to
cocoa growers throughout the world.[45]

Hershey's main rival, Frank Mars of Minneapolis, emerged a generation later when the young Mars added too much egg albumen to his

**Figure 4.4** An outdoor billboard from June 1923 advertising a popular combination candy bar. Note the continued appeal to the woman and the idea of her sharing her Oh Henry at parties, rather than as merely a personal indulgence. Library of Congress (LC-USZ62-83203).

employer's chewy nougat recipe, creating (by accident) a new taste with an airy nougat. Enrobed in chocolate, Mars created the Fat Emma bar, a name fortuitously changed in 1923 to the Milky Way. Mars's company really took off with its fabulously successful Snickers bar, unveiled in 1930. Frank's son Forrest was then sent to England, where he launched his own company in 1932. In 1941, the younger Mars teamed up with Bruce Murrie, son of Hershey company president William Murrie, to make a small pellet-like candy known as M&Ms, so named for Mars and Murrie.[46]

The Mars Company made its mark with combination bars that introduced a profile of flavors. In the Snickers bar, for example, an outer layer of semisweet chocolate was followed by salty peanuts, caramel, and a center of nougat. The combination bar was not a Mars invention; peanuts had been embedded in molasses from the 1830s, and similar combinations were key to the bonbon's success. But the idea of joining contrasting tastes went down market when a German immigrant by the name of Frederick Rueckheim combined sweet and salty in a popcorn-peanuts-molasses treat for the Chicago World's Fair of 1893, calling it (three years later) Cracker Jack. In 1912, Rueckheim introduced a toy premium to the box festooned with the famous sailor boy and dog trademark. Seemingly a simple snack appealing primarily to children and to baseball fans ("Take Me Out to the Ball Game"), Cracker Jack was innovative in combining contrasting flavors and textures.[47]

A similar approach was taken in 1912, when Standard Candy of Nashville introduced its Goo Goo Cluster made of caramel, marshmallow, chocolate, and peanuts. Otto Schnering's Curtiss Candy Company in 1920 started marketing the Baby Ruth, a peanut-chocolate-nougat bar. Like most other successes, Baby Ruths were advertised heavily in broadly circulating periodicals like the *Saturday Evening Post* but also in the *Open Road for Boys*. Many other combination bars from this era will still be remembered; these include George Williamson's Oh Henry (1920), a peanut, caramel, and fudge mix; California-based Peter Paul's Mounds (1922), combining coconut with semisweet dark chocolate; and Pittsburg's Clark Bar (1917), a meld of salty peanuts with milk chocolate. Hershey then followed up with Mr. Goodbar (1925), a chocolate bar with embedded peanut bits, and the cheaper and less memorable Krackel (1938) containing crispy rice. A more successful innovation was Reese's Peanut Butter Cup with crenulated chocolate enrobing, introduced in 1941 by a former Hershey employee, H. B. Reese.[48]

Combination candy bars offered a progression or blend of taste sensations, often based on sensory opposites — salt and sweet, hard and soft, rough and smooth, with contrasts also between different parts of the actual candy (the outer layers vs. the center, for example). The hope was to appeal to a more sophisticated taste by delivering more than just a sugar high. But we probably shouldn't make too strong a case for the refinement of chocolate bar aesthetics. From the beginning these were essentially calorie packs advertised as "pick-me-ups" and substitutes for regular meals. Snickers and other combo bars were promoted as alternatives to the midday meal for people presumably too busy even for soup and a sandwich. And plenty of calories were provided. In the early years many of these bars — like the Goo Goo Cluster — weighed upward of a quarter of a pound, more than twice the size (and calories) of today's typical candy bars. The Hershey's chocolate bar offered in the American Field Rations soldiers fighting in the Second World War delivered six hundred calories of energy.[49]

Sizes of candy bars have declined substantially ever since, thanks partly to manufacturers who've kept costs down by diminishing sizes.[50] Health-conscious consumers, though, have come to recognize even smaller bars as "empty calories" and "junk food"—expressions that did not even exist prior to the Second World War.[51] A great deal of attention has been focused on the bodily impact of such hyperfoods, though it is probably also worth taking at least some note of the isolating impacts of such "treats." Candies of this sort are now very often eaten alone, while driving, for example, or between meals as a quick fix to save on time that would be lost sharing a meal with others. Available virtually any hour of the day at many types of stores and vending machines, these ready-to-eat foods led to the modern habit of grazing, contributing to obesity and the decline of the sociable "breaking of bread" at meals.[52]

## You Scream, I Scream

No discussion of sugared snacks would be complete without some mention of ice cream, yet another packaged delight that offers a variety of textures and sensations, including conveyance of a sense of "nature defied." The sweetness of (plant) fruits is combined with the fat of an animal, but here also we have a food, or at least a quasifood and marvel, that stands outside seasons, a chill in the midst of summer. The novelty is profound but also easy to overlook, given how common it has become in our daily lives. Ice cream is now a $14 billion business in the United States, and Americans eat about three million *tons* every year.

Of course the idea of a chilled confection itself has a hoary history: the ancient rich and powerful kept ice houses for cooling summertime dishes of honey, fruit, and even nuts and cereals, and despite warnings from the likes of Hippocrates against consuming iced waters in summer (given the humoral haywire), aristocrats and kings chilled wine with snow and ice anyway. In the fourth century BCE Alexander the Great had thirty trenches filled with snow from the mountains (and covered

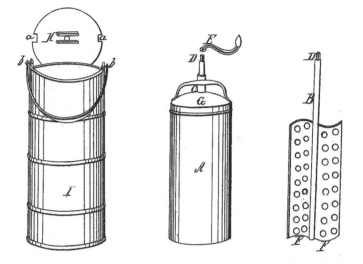

**Figure 4.5** A patent for a simple but popular style of device for homemade ice cream, 1843. U.S. Patent Office (no. 3,254).

with branches) for making cool drinks for friends in summer. Flavored ices or sorbets were served by the Mughal emperors of India, luxuries that would pass into Italy and France in the sixteenth century. True ice cream debuted in England in the 1740s and was served in the pleasure gardens in and around London from the 1780s, accompanying summer concerts. Like many other delicacies, ice cream passed from aristocratic to middle- and then lower-class tastes, but only after crucial inventions from the era of the packaged pleasure revolution.

Patents for hand-cranked ice-cream makers were filed in nineteenth-century America, including Nancy M. Johnson's 1843 maker, consisting of a tall tub of ice housing a cylinder with a close-fitting lid and an internal "dasher" (scraper). When turned by an external crank, the dasher would mix the freezing cream, sugar, and flavoring within the cylinder while also scraping the hardening concoction off the cylinder's inner wall. A five-minute ice cream freezer (1856) was popular for many years, though commercial prospects were limited by the requisite hand-cranking. Most ice cream businesses were small, requiring many hours of hard manual work, until electric power provided relief in the 1890s. Itinerant ice cream vendors had nonetheless first appeared in New York and other cities in the late 1830s, offering homemade ice cream—often of dubious quality. By midcentury ice cream was also being served in soda fountains, whose owners tried to outlaw street sellers. And ice cream became more sanitary—or at least less messy—when the common street practice of dispensing in a glass dish was gradually replaced with the unveiling of the "ice cream cone" at the St. Louis World's Fair of 1904. Consumers also welcomed the convenience of finger-friendly treats like the Eskimo Pie, a chocolate-covered ice cream bar introduced in 1920. Even more manageable was the ice cream sandwich, with cookie wafers substituted for bread. Lollipops must have helped inspire the Good Humor ice cream bar on a stick (and the Popsicle), which came onto the market in 1920 (and 1924).[53]

As with other sweets, ice cream was long identified with the taste and leisure cultures of women and children. By the 1860s it was being sold in bright and airy soda fountains and ice-cream parlors, contrasting sharply with the dark and alcoholic saloons of lesser (and more masculine) repute. And ice-cream parlors were often located in or near the new and exciting department stores. Outfitted cheerfully with potted plants, sparkling clean counters, and soda "jerks" nattily dressed in spotless white uniforms, these parlors served as a female-friendly retreat from the smell and dangers of beer and whiskey. And again like other sweets, ice cream could be combined with diverse flavors in a variety of forms and textures. It could even be blended into flavored soda water, a fashion popular from the 1870s as the ice-cream soda. James Tufts's patent of the Lightning Shaker in 1888 introduced the era of the milk shake and, with eponymous flavoring, the malt. These were widely regarded as refreshing alternatives to alcohol—though not without some curious resistance. In the early 1900s, when some states banned soda along with liquor and beer on Sunday (apparently unable to distinguish soft drinks from hard), ice-cream parlors came up with the sundae, a high-energy concoction of ice cream and flavored sugar syrup.[54]

Ice cream was somewhat different from candy, though, in that it could always be portrayed as more foodlike than a candy snack. Consumers were getting dairy and eggs, after all, and even the cold could be taken as a relief from oppressive summer heat. Ice cream was similar to cake in this healthful respect, and the two were quickly paired in much the same fashion as nuts or coconut with chocolate. In the second half of the nineteenth century, ice cream and cake would become central fixtures in birthday parties and weddings.[55]

Parties featuring flavored ice were clearly already taking place by the eighteenth century, however, and that is one interesting contrast with other forms of sugar consumption: ice cream tended to have a more public and social character. Ice cream making was often a communal event, produced locally and noncommercially, with everyone taking a hand in the cranking. Eating the stuff was also inherently more public—and episodic—because of its perishability: it was virtually impossible to store for any length of time (until domestic freezers) and could not be concealed in one's palm or pocket. All of this contrasts with candy, which generally speaking had a long shelf life, could be squirreled away for eating in private, and could easily be concealed in the hand or pocket. Ice cream eating would eventually take on more furtive, private characteristics—whence bingeing—with the invention of domestic freezers, but those would not become widely available in private households (as opposed to commercial establishments) until the middle decades of the twentieth century. Bingeing in this sense has socio-technical roots, having as much to do with revolutions in domestic appliances as with personal sins of sloth and gluttony.

## Chilled Fizzy Water

Fermented beverages have been favorite table drinks for millennia in most parts of the world. For Europeans in the seventeenth century, malted grain fermented in water (commonly called ale) vied with fermented fruits (wine) along with distilled fermented grain, fruit, and molasses (rum). American colonists drank mild beer from the age of seven with meals, including breakfast. One of the first things the Pilgrims thought of shipping from England was malted (partially germinated) barley for beer, but in a pinch they also made "beer" from readily available corn; roots and fruits like parsnips, pumpkins, and tomatoes; or even spinach and dandelions. The ideal in colonial America was a mild "refreshment," meaning drink that would not lead to rowdy inebriation. Overindulgence was strongly discouraged and public drunkenness in New York could lead to three quarts of salted water laced with lamp oil being poured down the throat of the offender. But not even the gains of the temperance movements from the 1830s or even prohibition efforts (that made the state of Maine "dry" as early as 1846) could stamp out drinking. It was not prohibiting alcohol but developing alternative "refreshments" and making them routine that would have more success against booze in the nineteenth century.[56]

Of course there were several alternatives for the title of America's routine beverage, including several normally served warm, such as coffee, tea, and chocolate. But a chilled drink competitor emerged in the middle of the century in the curious form of carbonated water. Our common term "soda" was coined in the 1790s (from the sodium carbonates used in its manufacture) but comparable drinks go back centuries, to whoever first drank bubbly mineral waters containing natural carbon dioxide. Hippocrates circa 400 BCE chronicled the virtues of carbonated water, and as the Roman Empire expanded, many of the mineral springs of England, Germany, Belgium, and Italy attracted Romans convinced of their medicinal powers. Springs of this sort were mostly used for curative bathing rather than drinking. Faith in the healing powers of mineral water migrated to colonial North America in the eighteenth century, and by the early 1800s the health-conscious well-to-do had started gathering around the springs of Saratoga, New York (and later White Sulphur Springs in West Virginia) to drink foul-tasting mineral waters. Enlightened Americans, including Thomas Jefferson and Benjamin Rush, shared and wrote about this enthusiasm.

Efforts to artificially carbonize water for drinking date from 1685, but only in 1772 did the English chemist Joseph Priestley, best known today for discovering oxygen and pencil erasers, succeed. Priestley suspended a bowl of water over a beer vat in a Leeds brewery, allowing the "fixed air" (that is, carbon dioxide) from the fermenting beer to percolate into the water. Later he discovered that dripping sulfuric acid onto chalk created carbon dioxide that could then be dissolved into water.[57] Priestley sold his carbonated water to the Royal Navy as a cure for sea scurvy, in an era prior to the recognition of vitamin C deficiency (British sailors would later be called "limeys," acknowledging the more successful preventative of lime juice). However useless against scurvy, carbonated water did seem to be an intriguing drink. Nicholas Paul and Jacob Schweppes began manufacturing soda water in Geneva and London in the 1780s, which is also about when Joseph Bramah, an English locksmith and inventor of the flush toilet, developed a continuous process for carbonating and filling bottles that could hold a certain level of pressure.[58]

By 1806, manufactured soda water was available in the United States both in bottle and in draught form. Bottles were still rare and expensive, however, so for nearly a century soda water would come mostly from taps attached to pressured canisters via metal (and later rubber) tubes. The first person to open a commercial soda fountain in the States was Benjamin Silliman, a Yale chemist and mineralogist who in 1809 installed a hand-pumped setup at the famous Tontine Coffeehouse at the corner of Wall and Water streets in New York City. Silliman manufactured his carbonated water in the cellar, from where iron tubes led upstairs to a decorated manual pump from which customers could be served. Gooseneck fountains were installed in hotels and in urban health spas and, from 1825 onward, in drugstores. Over time, the fountain dispensers became progressively more elaborate, with the spigots set in urns or between ornate marble columns. Although drunk as a refreshment, once again many consumers regarded chilled carbonated water as medicinal. However, carbonated water was removed from the American book of pharmacopeia in 1831, signaling its transition from a therapeutic tonic to a pleasure drink. By this time soda water was being mixed with fruit and other flavored syrups, along with generous portions of sugar.[59] Soda thus joined candy as a new kind of manufactured delight, offering a distinctly invigorating hit of pleasure.

By the mid-nineteenth century, soda fountains featured an astonishing range of (mostly sweet) flavorings, forging the basis for a revolution in tastes that would rot millions of teeth and expand millions of waistlines. Extracts and distilled oils were easily blended with the bubbly yet neutral-tasting carbonated water. Berries and other perishable fruits were mashed and strained into syrups to which sugar was then added—both to sweeten and to preserve the flavors. Wild roots and herbs were also soaked in alcohol to draw out natural oils. Boiling

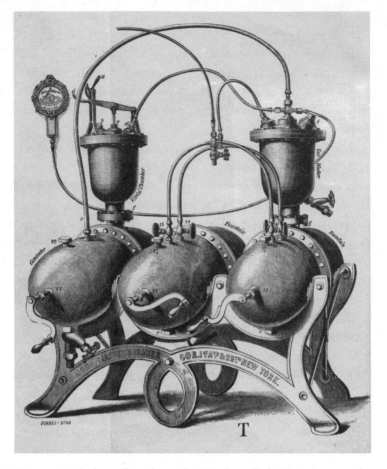

**Figure 4.6** A picture from the catalog, Matthews Soda Water Apparatus by John Matthews, 1871, showing a device designed to carbonate water in basements of soda fountains.

or distilling water-saturated plants produced still other syrups, with ribwort, nettle, birch bark, currant leaves, sassafras petals, and even dandelion added to the soda water repertory. Many of these roots and herbs had already been used in therapeutic regimens, and early soda sellers often made health claims for their concoctions.[60]

Where did such a varied range of flavors come from? Soda fountain innovators borrowed from folk recipes for rural American "small beers," nonalcoholic brews typically made from water, molasses, and yeast combined with additives of roots and herbs found near the farm house. Such was the origin of sarsaparilla and birch and root beer, many of which were modified from drinks developed first by Native American populations. Soda fountains would often modify the composition of some particular root or small beer, replacing, say, the molasses with cane sugar to increase its potency over the folk version. By the 1850s chemists were also creating artificial flavors — typically from compounds first discovered in the perfume industry — a first step to our modern world of the professional taste engineer, sensory scientist, and "flavorist."[61] Pierre Lacour, the French manufacturer of liquors and wines, published in 1853 a book of recipes for soda, including extracts from lemon, pineapple, orange, strawberry, vanilla, peach, grape, almond, spirit of roses, blackberry, mulberry, raspberry, neroli oil (bitter orange and orris root), and ginger. By 1865, there were 123 makers of soda syrups in the United States, offering a broad palette of flavors ranging from apple and gooseberry to melon, cherry, and allspice. By 1890, additional flavors had been added, including wintergreen, walnut cream, champagne currant, cranberry, checkerberry, chocolate, coffee, kola berry, blood orange, and banana. Stranger to our modern ear, perhaps, were sodas based on celery, crabapple, pepsin, orgeat (a barley-almond blend), syrup of violets, and calisaya — a concoction of cinnamon, cloves, nutmeg, ginger, and tolu (an aromatic balsam from a South American tree). The Gay Nineties lived up to its name with faddish new drinks like Cream Zoo, Choctow Sling, White City Dew, Suburban Featheredge, and Foam Fun.[62]

All this variety and novelty marked the soda fountain as a particularly effervescent site of the packaged pleasure — and discernibly gendered. By midcentury the soda fountain had become an acceptable public setting for female socializing. In lieu of the foam on the beer in male-dominated saloons, fountain drinks offered froth and sweetness for women, with chocolate and fruity flavors especially in demand. Most soda syrups were heavily sugared: typically five to six pounds per gallon. This too was gendered, since men tended to hang on to the notion that soda could cure stomach or other ailments — which is why adult males tended to favor bracing, even foul-tasting, sodas (or "tonics") flavored with chartreuse, quinine, and phosphors. But the soda fountain was predominantly a female place—from the consumer side at any rate—where a mind-boggling array of flavors could be chosen, all offering uplifting refreshment and a hit of sucrose.[63]

Adding to the sensation was the bracing chill itself, another "break" with nature. Early soda had been served at room temperature, but this changed dramatically when James Tufts, a druggist from Somerville, Massachusetts, invented the "Arctic" cooler in 1863. This device chilled the carbonated water on its way to the syrup flavoring, making possible a cold, refreshing beverage even during the heat of summer. Later, in 1891, Tufts cofounded the American Soda Fountain Company to supply equipment for retail dispensers. Refrigeration improved with the introduction in 1892 of liquid ammonia in the cooling condensers of refrigerators, making it easier to preserve ice cream and other "iced" products. The appeal of chilled drinks expanded also with the invention of the ice-cream soda in the 1870s. The cooling sensation of a frosty glass of ice cream on the hands and fizzy fluid drawn into the parched mouth via a straw, in our time utterly common, was once a stunning experience of refreshment. And yet another example of the novel sensory delights ushered in with the packaged pleasure revolution.

The soda fountain experience itself, though, was a novel sensorium and new kind of social site. Its ideal look, as expounded by James Tufts himself, was a polished metal counter with clean white towels keeping it shiny and spotless, attended by a well-mannered young man (not a boy) dressed in a white coat and exuding "tact and limitless good nature." The experience was to be one of safety and sobriety—and the Victorian feminine appeal is obvious,[64] insofar as soda fountains offered a new kind of pleasure that energized and refreshed without the buzz or disrepute of alcohol (though some soda drinks actually contained small quantities of wine). Subsequent to the epidemic of temptation that had swept the country in the 1820s and 1830s in the form of cheap booze and gambling and associated sins, soda—like candy and chocolate— offered a more upstanding treat.[65]

But the nineteenth-century soda fountain maintained a mystique that went beyond its promise of exotic flavors, fizz, and sugar highs. With long-standing links to patent medicine cures and their typi-

**Figure 4.7** A typical spotless soda fountain, circa 1900, the focal point of the drug store. Detroit Publishing, Library of Congress (LC–D417–404).

cal housing inside drug stores, soda fountains promised relief from chronic fatigue and other ill-defined maladies. Consider the example of Charles Hires and his famous root beer. As early as 1800, pharmacies had sold baskets of roots and berries for "root beer," for which there were not yet standardized recipes or commercial brands. In 1870, Hires, an enterprising twenty-year-old druggist from Philadelphia, packaged these concoctions for sale for convenient home brewing. His mix of sixteen roots, herbs, and berries, including wintergreen, licorice, birch bark, sassafras, juniper, and vanilla, but also pipsissewa, spikenard, deer tongue, and dog grass, when mixed with yeast and water, made a soda drink that he first called root tea—to avoid the wrath of prohibitionists. Hires then in 1876 sold a still more convenient 25¢ pack of powdered mix that made five gallons of root beer when combined with water, yeast (for the fizz), and sugar. Hires was only packaging ingredients that were commonly known to rural Americans (even George Washington had had his own recipe containing bran and molasses). But Hires went further by drawing on another old tradition, claiming that his root "tea" was medicinal. The man even advertised in church and youth magazines, hoping to remain on the right side of the temperance movement. But when he dared to call his soft drink "root beer," the Women's Christian Temperance Union condemned it in 1895 as leading the young astray. The company fought back, insisting in *Ladies' Home Journal*, for example, that Hires's root beer was not just a temperance drink, but also a purifier of the blood and an overall health tonic (see figure 2.4, a Hires trade card).[66]

Other soda makers followed suit, making health claims to win over a wary middle class. (Recall that we are talking about the heyday of patent medicines, when nothing in the way of tests was required to distinguish legitimate from bogus nostrums.) In 1885, Augustin Thompson, a New England pharmacist, mixed an elixir of sugar, cinchona, sassafras, caramel, and gentian, a flowering plant from the Pyrenees and the Alps, to which he gave the grand brand name of Moxie Nerve Food. Thompson advertised his wonder drug as an appetite and strength builder, promising also to cure a suspiciously broad set of ailments ranging from nervous exhaustion, paralysis, and "softening of the brain" to insanity, impotence, and imbecility. Thompson later added soda water to his concoction and sold it in bottles and at fountains. Moxie soda was very much an acquired taste, however, causing Thompson to offer cheap Moxie lollipops to win over kids to its bittersweet taste. In the 1920s, "moxie" became slang for "courage" or "nerve" when the beverage was advertised with the slogan, "What this Country needs is plenty of Moxie."[67]

Dr Pepper was another soft drink promising both health and vitality—with novel ingredients added to give a zesty zing to the drink. Again a creation of a druggist, Charles Alderton, in the dusty town of Waco, Texas in 1885, Dr Pepper was a secret combination of fruit flavors and mysterious spices from as far away as Madagascar, all dissolved in distilled sparkling water. Hoping to associate his drink with good health, Alderton named it Dr Pepper. Sugar, of course, was the principal ingredient, but it also contained another compound that must have helped its success—caffeine, first added to the formula in 1917. Dr Pepper was sold as a cure for alcoholism, smoking, indigestion, age, nerves, and exhaustion—and consumers were urged to not be shy about using it liberally. One ambitious ad advised that the drink be taken regularly, essentially as medicine, at 10 a.m., 2 p.m., and 4 p.m. to avoid daily fatigue.[68] Soda consumption grew rapidly in the final decades of the nineteenth century and the first decades of the twentieth, with a certain consolidation of brands as in other consumer realms (cigarettes, for example). Countless flavors were explored, but two would come to dominate, lasting even into the present. The first was ginger ale, an old Asian standard whose makers included Clicquot Club (1881), White Rock (1883), and Canada Dry (1904). Heavily advertised, this drink pushed aside many more exotic fruit and root flavors in the first decade of the twentieth century. In 1921 alone, Clicquot Club ginger ale was advertised in twenty-five national magazines touting its cool refreshment in the summer months. Clicquot Club was promoted as not only thirst quenching but "brain cleaning."[69]

The second flavor to consolidate this variety was *cola*, a mix of fruit essences to which the kola bean was added. Unlike the ginger ales, colas offered more than just sweetened, flavored fizz water. Kola beans also contained the alkaloid caffeine and came with an exotic aura, having long been used in Ghana, Brazil, and the West Indies as a stimulant, a cure for hangovers, and an aphrodisiac. Moreover, the leading cola manufacturer, Coca-Cola, had added another ingredient that, for a time at least, would pack still more punch: five ounces of extract of coca leaf per gallon of syrup.

Coca-Cola's creator was John Pemberton, a prolific inventor of patent medicines working out of his Atlanta, Georgia, drug store. By the time he came out with Coca-Cola in 1886, this Confederate veteran had already offered his customers Extract of Styllingia (a blood purifier), Indian Queen Hair Dye, and Triplex Liver Pills and Prescription. Pemberton had earlier also manufactured a "French Wine of Coca," which he had touted as nerve tonic, being also a cure for dyspepsia, mental exhaustion, gastric irritation, constipation, headaches, and even opium addiction. Pemberton's coca wine was largely a copy of Vin Mariani, a popular European patent medicine endorsed by Thomas

Edison, J. P. Sousa (brass band leader), Lillian Russell (the singer), Wild Bill Cody (Buffalo Bill), Sarah Bernhardt (the actress), and even Pope Leo XIII. Pemberton's coca wine contained a coca leaf extract whose chief alkaloid, cocaine, was favored by Sir Robert Chistison, president of the British Medical Association, and by Sigmund Freud, who considered it a useful stimulant and an antidote to depression. President Ulysses S. Grant used cocaine to relieve pain from the cigar-induced throat cancer that ultimately claimed his life, and many druggists sold the alkaloid over the counter, no questions asked.[70] Cocaine was not restricted in the United States until the Harrison Act of 1914, spurred partly by fears that the drug was impairing people's ability to work. The price of cocaine by this time had dropped to such a point that it was sometimes cheaper even than alcohol, and prohibition was further fueled by racist fears that the alkaloid was being used by lascivious "negro cocaine fiends."[71]

Atlanta went "dry" for a couple of years in the 1880s, causing Pemberton to try to placate the temperance crowd by eliminating the wine—while keeping in the coca.[72] But as cocaine lost its glamour as a wonder drug, Pemberton and then his successors reduced its presence in Coca-Cola, eliminating it altogether in 1903 except for a trace. As for the kola nut, the company found a cheaper substitute in caffeine. But the most important ingredient in all such drinks was always sugar. Coke contained five teaspoons in an eight ounce bottle—a not-uncommon proportion for soft drinks from this era. Indeed, like most soft drinks, 99 percent was just sugar and water, with the remaining one percent consisting (in Coke's case, by 1917) of caramel for coloring, glycerin for body, phosphoric acid for zip, and an assortment of essential oils from lemon, lime, and orange, along with vanilla, neroli oil and nutmeg. Silly talk of a secret formula passed into company folklore and propaganda, but like virtually all other sodas or soft drinks, Coke was basically just flavored, caffeinated sugar water.[73] And the "secret formula"—such as it is, in several different versions—can now be found on Wikipedia.

Coca-Cola's success, of course, spawned numerous imitations. Pepsi Cola, introduced by Caleb D. Bradham from New Bern, North Carolina, in 1898, was only one of many. Like other successful packaged pleasures, Pepsi prevailed because it was heavily advertised, beginning in 1903. Bottling followed in 1904, with 280 licensed bottlers by 1920. Bradham claimed early on that Pepsi was a health drink (think about the name) with "none of the stronger narcotics" of Coke. In any case, it was sweeter. Pepsi gave Coke a good run for its money in the first decade of the twentieth century, but Bradham was forced to sell out in 1922 from having overspeculated in sugar during the First World War (prices dropped precipitously in the aftermath). The company dragged along with other owners until a rebound in 1936, when Pepsi offered a 12-ounce bottle for a nickel, the same price as Coke's six ounce bottle. The radio jingle drilled home the point: "Pepsi Cola hits the spot / Twelve full ounces, that's a lot / Twice as much for a nickel, too / Pepsi Cola is the drink for you."[74]

Americans' taste for soda seemed to know no limits, however, judging from the success of several new upstarts entering the market: Orange Crush in 1916, 7Up in 1920, and Royal Crown Cola in 1934. Soda became an on-the-go stimulant available almost anytime and anywhere. But there was a cost to this effervescence: a dramatic reduction of choice of flavors. The traditional soda fountain, with its combination of exotic and local flavors, gave way to the uniform power walls of franchise brands pushed by national ad campaigns and deals with retailers to shelve favored brands. Despite the long endurance of regional brands, only a select few brands survived mass-produced bottling. Hires came to dominate in root beer, Clicquot Club in ginger ale, and Coke and Pepsi among the three hundred-odd extant cola brands as of 1917. The foul-tasting Moxie found loyal followers into the 1920s, but only thanks to its advertising clout. Maybe celery soda didn't taste that great, but millions of Americans seem to have traded in a bouquet of flavors for the pride and convenience of sharing in the same small handful of beverages, mostly laden with caffeine and sugar and few traces of the original tastes of the fruits and nuts of nature.[75]

Coke and Pepsi for a time even came to dominate the world, dividing the globe in much the same ways as did Spain and Portugal in the sixteenth century. Coke got Europe, Pepsi got the Indian subcontinent. The (perhaps apocryphal) story now circulates that of the sixty-four fluid ounces of liquid drunk daily by humans throughout the world, only two of these ounces, on average, is Coca-Cola. The remaining sixty-two is the hoped-for market potential, room for expansion.

## Superfoods as Junk Foods

There is a kind of irony in the fact that junk foods, especially those based on refined sugar, offered to the masses food and drink previously available (if at all) only to the rich. These junk superfoods at first supplemented traditional diets, providing convenient and high-energy quick hits for working people cut off from the land, offering cheap calories with little or no time required for meal preparation: you just tore off the wrapper or popped off the top. At the beginning of the nineteenth century, sweetened fruit preserves could be stored without refrigeration while jam on white bread with sugared tea got manual laborers moving in the morning. By the end of that century, a Coke and a quarter-pound candy bar for some at least would replace the traditional midday family meal (or prepared lunch box).

Our story, then, is at once a tale of sensorial enrichment and nutritional impoverishment. In its vast variety of textures, tastes, and forms, sugar, with the help of concentrated flavor or milk, egg whites, cocoa, nuts, fruit, and especially fizzy water, vastly expanded the palette of industrializing people. But from a nutritional or health point of view, sugar-based foods provided little more than shots of fast-acting energy. Sensory enrichment, bodily impoverishment. In candy, chocolate, and soda pop, sugar offered a classic trick on the senses: taste and variety hiding very simple and inadequate nutrition. This new dietary regime democratized novelty, but at what cost?[76]

The rise of candy bars and colas can also be seen as part of an assault on local and regional cuisine and family meals. Jell-O replaced local variations in pies and pastries for dessert, just as Coke prevailed over a broader array of local brewing and harvest cultures. At the same time, the junk food revolution—a key element in the packaged pleasure revolution—let the individual eater break from family and social traditions. As historian Sidney Mintz notes, packaged sugar products made it possible "for everyone to eat exactly what he or she wants to eat, in exactly the quantities and under exactly the circumstances (time, place, occasion) he or she prefers." The chilled Coke beverage provided an instant gulp of personal satisfaction. Cheap candy also liberated children from their elders, just as fast food gave immigrants a path to assimilation (and escape from ethnic traditions). Chocolate, ice cream, and soda also offered women a respite from the confines of respectability in a form of acceptable dietary sensuality. But junk foods also tended to undermine the "companionship" of breaking bread with others. The larger result was the habit of focusing on "consumption itself—not about the circumstances that led them to consume." Few eaters probably thought about where the sugar came from, and not enough worried about long-term impacts on human health and sociability.[77]

All of this fit into the proverbial faster pace of industrial life, of course, whence "fast food." Many consumers of candy and soda worked more intense or longer hours, with ever-longer commutes to work than their rural predecessors. Faster life encouraged faster foods, even as these new consumers of packaged pleasures bathed in a culture insisting on the supreme value of individual choice, even if this meant only when and where to consume some standardized brand of junk food. The cheap, sweet package certainly met needs, but it also confined their expression.

From a larger historical point of view, the impact of fast food has been to corrupt the human body and human political institutions—insofar as the hyperfood industry now wields substantial political power. As recently as the early 1970s only 14 percent of Americans met the medical definition of obesity. By 2010, that rate had climbed to 35 percent.[78] History is often about how the world we know came into being, and what has been lost along the way. If we are moving into a world of epidemic obesity and diabetes, this is partly the result of technologies and corporate powers tracing back to innovations of the nineteenth century. Understanding how this came to be, may help us to better understand how it might come undone.

# Portable Packets of Sound

*The Birth of the Phonograph and Record*

Some of us will recall the Hans Christian Andersen fairy tale about the beautiful mermaid who strikes a devil's bargain, giving up her lovely voice to a sea witch in exchange for legs, allowing her to win the heart of her prince charming. At first she hesitates: "But if you take away my voice, what is left for me?"[1] But then she agrees. The story has a predictably happy ending (though in the Andersen, as opposed to the Disney version, the mermaid doesn't win her prince but rather ascends to the ranks of the Daughters of the Air). The undeveloped but intriguing question is: why does the witch long to "own" the mermaid's voice? Or for that matter, why would anyone desire to possess and control any voice or sound?

One explanation would be the mystical power of the human voice. In folklore and popular imagination, the voice is the heart and, literally, the soul of the person. It is the *word* of God (*logos*) that is captured in the Bible, not some divine law inscribed in nature. Despite Plato's fears that music subverted rational thought, the ancients believed music to be a gift from the gods. And the power to preserve sound was one of the qualities that distinguished gods from mere mortals. Controlling the voice would be an ultimate power: whoever possessed the divine voice controlled the deity. Egyptian priests at Thebes (ca. 1490 BCE) empowered a statue of Memnon with a "voice" using hidden air chambers; later, Greek oracles enchanted and deceived their devotees by inducing statues of the gods to "speak."[2]

This quasi-supernatural power obsessed early philosophers in their efforts to create machines that could reproduce the human voice. Roger Bacon in thirteenth-century England endeavored to construct a mechanical talking head made from brass. Giambattista della Porta in Italy in 1589 fantasized about preserving spoken words inside leaden pipes, captured in such a way as to burst forth when their covers were removed. Cyrano de Bergerac in the seventeenth century wrote of a time when people would be free to listen to a book while walking rather than being tied to the burden or skill of reading. Similar dreams cropped up again and again until the 1840s, when the German American Joseph Faber tried to produce artificial speech from a vibrating ivory reed and a rubber tongue and lips controlled by a keyboard.[3]

Of course, transcribing oral expression by means of abstract symbols has a much older history. Writing is essentially a form of bottled speech, stored in the form of a visual mnemonic code. Writing allows us to preserve and transmit spoken words and sound across vast stretches of time and space, doing basically for thought what pottery does for food and drink. Musical notation was a later development: there are hints of recorded tones in ancient Greece, but schemes ancestral to those used today date only from the European Middle Ages. A written twelve-note semitone scale dates from the tenth century, the four-line grid for recording notes appeared by 1250, and signs for rhythm by 1316. Notation made transgenerational music easier, along with the world's first possibility of an enduring "accumulation" of music and musical traditions.

But the abstract and symbolic exercise of writing and reading words and music is nothing like the magic of recorded speech and music. The impermanence of speech and music before recording, its incorporeal and subjective nature, meant that it could not be easily collected or ultimately possessed. This, according to music historian Evan Eisenberg, made music uniquely precious in the eyes of nineteenth-century romantics. Sheet music might fix the score and reduce the need to memorize, but the performance itself—the essence—remained ephemeral, unpossessed.[4] All that changed in 1877, when Thomas Alva Edison created the world's first phonograph, trapping (eventually) even the most inspired of sounds—initially (and ironically, if fittingly) for business rather than for art.

## Capturing Sound for Capitalism

The breakthrough in the centuries-old quest for mastering the voice came, as many things do, by breaking from old expectations of how it might be done. While Faber and his predecessors had attempted to imitate the human voice by mechanically recreating the human mouth and vocal cords, the phonograph offered a new and simpler way of reproducing speech—by imitating not the mouth but, oddly enough, the ear. At the same time, recording sound emerged not directly from the dream of transcending ephemerality but rather from the more practical goal of perfecting business technology. This was a classic case of technology begetting new technology: when the railroad "annihilated" distance in transporting people and goods, an even more rapid means of communicating over long distances was required, even if only to announce the arrival of people and goods. This led to the telegraph and the telephone, which in turn required new means of transmitting (and eventually recording) messages formerly sent by mail. Simply put, the increased pace of business communication produced the phonograph.

Thomas Edison, an entrepreneurial inventor of machines for counting votes and recording sales on stock exchanges, in February 1876 saw a need for improving the transfer and storage of telegraph messages. This coincided with Alexander Graham Bell's invention of the telephone, a contraption modeled after the eardrum, expressing his earlier fascination with finding a way to communicate with the deaf. Bell had linked an iron diaphragm to an electromagnet, transforming sound waves into a modulating electrical current which, after its journey over wire, became sound waves again in a second vibrating diaphragm, the receiver. Edison understood the significance of Bell's invention and improved on it with his carbon diaphragm transmitter in 1876. From the telephone in fact came the phonograph, in that both were built on the ear-like diaphragm. Even so, it was only by accident that the "Wizard of Menlo Park" shifted his attention from the telegram repeater to recording voice. While experimenting with recording and transmitting Morse code using paper indentations on a cylinder, Edison found that "when the cylinder carrying the indented paper was turned with great swiftness, it gave off a humming noise from the indentations. . . . This led me to try fitting a diaphragm to the machine, which would receive the vibrations or sound-waves made by my voice when I talked to it, and register these vibrations upon an impressible material placed on the cylinder."[5]

Edison found that when a stylus was placed in the just-recorded groove and the cylinder turned on, the diaphragm vibrated, reproducing his just-spoken words (albeit faintly). He thereupon instructed his machinist, John Kruesi, to make what would become the world's first phonograph, consisting of a cylinder and a diaphragm/stylus (the "reproducer"). When the cylinder was turned with a feed screw and crank, the reproducer moved in a spiral groove across a tin-foil sheet wrapped around the cylinder. During recording, the sound waves jostled the diaphragm up and down, causing the stylus to make a spiral string of indentations along the groove. When later drawn back over these same indentations, the stylus reproduced the identical movement of the diaphragm, regenerating the previously recorded sound. With this simple device for copying and replaying short voice messages, Edison hoped to replace the tele*graph* repeater with a tele*phone* repeater, allowing the transmission of a recorded voice rather than a coded message. In November 1877 he demonstrated his device to the editorial staff of *Scientific American*, who quickly recognized its significance and agreed to publish it.[6]

Edison was not the first to conceive of recording voice in this way. A Frenchman by the name of Charles Cros had published a description of a similar device (using a disc rather than cylinder) in April 1877 but had found no backers for his invention. So Edison's patent, filed in January 1878, won the day. Savvy in the world of business, the young inventor immediately set up the Edison Speaking Phonograph Company and hired James Redpath of Boston to publicize his device on a national tour. Edison also sold showmen rights to demonstrate the

**Figure 5.1** A young Thomas Edison with the first edition of his cylinder phonograph, 1878. Library of Congress (LC-USZ62-98128).

phonograph, circus-like, on a stage where enthralled crowds would hear voices recorded in various languages and even dogs barking.[7]

For all its genius, Edison's phonograph was really just a composite of already well-known technologies: the stylus, the diaphragm, the cylinder, and the feed screw all had long histories. Indeed, Edison's critical innovation was arguably just the insight that sound could be recorded. The technical means used to produce the device were hardly even on the forefront of technology; the phonograph was a kind of throwback to an older generation of tinkering. Edison was perfectly familiar with the latest in telegraph and telephone technology, but his phonograph transmitted sounds by purely mechanical means — with no assistance from electricity or magnetism (as later would be the case).[8]

Edison was prolific and easily distracted, however, and within a year had lost interest in the phonograph, shifting his attention to his greatest contribution, the electric light, which he patented in 1879. We should also not forget that the sound quality of his early phonograph was poor, and repeat playing left much to be desired. The tin-foil cylinders were ruined after only two or three plays, and irregularity in the speed of cranking created serious distortions in the sound. Time would eventually wear on the device, and today we might even smile at the crudity of such reproductions.[9] Crucial for us to realize, however, is the revolutionary nature of such an invention, even if originally imperfect. Here in embryo was a machine that could make a "written" record of sound — hence the name "phonograph" (literally, "sound writer"). More striking still was the fact that the new device "read" back that record with no human effort, apart from turning a little crank. *Scientific American* in December 1877 commented on how Edison's invention "saves us that trouble by literally making it read itself" in playback. The first impression was astonishment at the new tool's simplicity. Contemporary media were impressed, as when *Harper's Weekly* on March 30, 1878, declared: "Herein the phonograph seems actually to have got ahead of that other marvelous construction, the human body." Onlookers (and listeners) marveled that the contraption talked without lips or tongue; others remarked on how a metal diaphragm had been made to listen and to speak, simplifying or even besting human nature. Much was made of the fact that the phonograph could preserve the living voice and, in a manner of speaking, raise the dead. A letter to *Harper's* in November 1877 waxed ecstatic over how, with Edison's device, "speech has become, as it were, immortal." Near-magical powers were granted to this wondrous novelty, powers that we today unthinkingly take for granted.[10]

The phonograph was also heralded for the control it seemed to give its owners over other people. *Harper's* in March 1878 declared the phonograph to be like an obedient servant, insofar as it "does not speak unless spoken to." In the wrong hands, though, it could just as easily turn traitorous since, as this same author put it: "this little instrument records the utterance of the human voice, and like a faithless confidant repeats every secret confided to it whenever requested to do so." Edison himself gleefully wrote about the power of his cylinders to capture and retain "all manner of sound-waves heretofore designated as 'fugitive,' . . . without the presence or consent of the original source, and after the lapse of any period of time,"[11] as if the fantasy of the sea witch had come true: the possession of the human voice. Here in wax and iron was the triumph of that age-old human vanity, the hope to match or even to surpass nature, defeating the ephemerality of the senses or even the uniqueness of separate and distinct moments in time.

Despite such high-flowing rhetoric, Edison's original goal was actually more mundane. His particular hope was to find a more direct way of archiving and transmitting the human voice as an alternative to written records, mostly for business purposes. Looking back, it is strange how radically Edison underestimated the significance of his achievement in May of 1878, when he commented that the "main" use of the phonograph would be for dictation. An automatic sound recorder would allow the busy entrepreneur to record ideas at will without writing, enabling him to "busy himself about other matters." The device would also eliminate the "confidential clerk." Edison was already thinking about the kinds of clerical automation we today associate with the personal computer—including the automatic storage of voice and freedom from human intermediaries in recording, replaying, and editing ideas, information, and plans. He also envisioned other uses: "We will be able to preserve and hear again . . . a memorable speech, a worthy singer . . . the last words of a dying man . . . a distant parent, a lover, a mistress." He also saw an opportunity to preserve languages as currently spoken for posterity (notably those on the verge of extinction), but also to use recording for rote learning, and even for talking clocks and dolls.[12]

Curiously, Edison ranked the recording of music only fourth on his list of applications, after dictation, reading aloud to the blind, and the provision of lessons in spelling and language training, proposing not commercial music but that "a friend may in a morning call sing us a song which shall delight an evening company." Some of his contemporaries, however, went further. A *Harper's Weekly* article claimed that the great luminaries of the era might be made to "sing or speak once in any place, [and] their words and tones will be captured by the phonograph." The tin-foil record "will be electrotyped, and copies sold at so much a piece" so that "at any time" one could hear "the great ones of the earth discourse in our own parlors." Edison himself dreamed that the "actual voice of the newest prima donna or exquisite melodies of a Liszt, a Mozart or a Beethoven will be rendered in all their beauty" on the phonograph.[13]

Even a decade later, though, Edison still thought of his invention as primarily a device for generating what he called "phonograms," including "audible letters" and documents. The problem was that with the technology of these early years, these messages could be played back only a few times and were duplicated only with difficulty. Sounds could be preserved, but they could not yet be mass produced.[14]

## The Talking Machine Reinvented, 1888–1900

For Edison, the phonograph really was a diversion, part of his passing interest in the possibilities of a "repeating" telephone, a device that could record and store a particular voice message. And while Edison was turning to create a direct current electrical system to power his heated-filament light bulb, other inventors turned to improving the phonograph. One of these was the inventor of the diaphragm-based telephone, Alexander Graham Bell. In 1880 Bell won the Volta Prize of $10,000 for his invention of the telephone and used the money to set up a laboratory in Washington, D.C., to further his electrical and acoustical research. There, with his cousin Chichester Bell, a chemical engineer, and Charles Tainter, an instrument maker, Alexander Graham worked on a waxed cardboard cylinder to replace Edison's tin-foil record. On May 4, 1886, the Bell group patented an improved talking machine, dubbed the "graphophone." Shortly thereafter they asked Edison to join them, even offering to put their improvements under his name. In his typical style of cantankerous independence, the Menlo, New Jersey, inventor refused and returned to improving his old device with a hard wax cylinder and two sets of mandrels and diaphragms—one with a sharp stylus for recording and the other with a more rounded needle

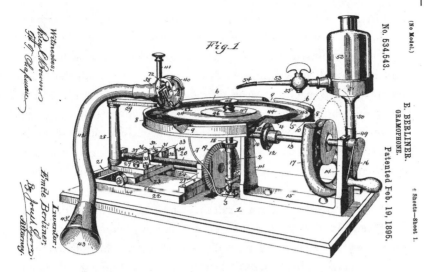

**Figure 5.2**  Emile Berliner's patent for the 1895 Gramophone, precursor of the Victor talking machine. Note the sound tube on the left. U.S. Patent Office (no. 534,643).

to play back. By June 1888, Edison was once again hoping to market his new and improved phonograph as a business machine, intended for recording dictation and as an answering service for use in hotels.[15]

All of these new talking machines used an etched, hard-wax cylinder to store the recorded sound — with one notable exception. In 1887, a German immigrant by the name of Emile Berliner came up with the idea of a "Gramophone," with the recording element consisting not of an elongated cylinder but rather a flat disc. So whereas both Edison's and Bell's needles had cut *vertical* grooves from the vibrating diaphragm, Berliner's cut *lateral* impressions into flat discs containing smooth grooves. Berliner's discs required a more elaborate setup for manufacture and use, making them less appropriate for home recording and playback, but they were also cheaper to manufacture, allowing prerecorded discs to be sold at somewhat lower costs than Edison's cylinders.

Berliner's prerecorded discs marked the beginning of a professionally organized sound-recording industry. Lacking funds, however, Berliner's discs were at first used in novelty dolls (to make them "talk"). And in the meantime Jesse Lippincott, an investor in the graphophone, acquired the rights also to distribute Edison's phonograph through the North American Phonograph Company, making the cylinder record the market standard. By 1890 most American cities had at least one franchise supplied by Lippincott's company, leasing both Bell/Tainter and Edison machines. And although enthusiasts still hoped that the phonograph would record famous voices and preserve dying languages, with some even wanting to record the wailings of ghosts in the séances then fashionable, Lippincott and his franchisees concentrated on leasing phonographs to businesses, essentially as secretarial aids and labor-saving devices.[16]

Despite Edison's energetic salesmanship, businessmen found his improved phonograph hard to use in recording, and harder still to understand in playback. As a result, most of his franchises failed fairly quickly, along with the parent North American Phonograph Company. One notable exception was the Columbia Phonograph franchise located in Washington, D.C., the embryo of what we now know as Columbia Records. By 1893 Columbia, an offspring of Lippincott's company, had replaced the Bell/Tainter enterprise as Edison's chief competitor in cylinder phonographs.[17]

In the 1890s, phonograph companies shifted from business to entertainment uses for recording sound, realizing that a major change had occurred since Edison first invented the phonograph in 1877. A mass consumer culture was starting to emerge in the United States, with burgeoning markets not just for goods and services judged "necessities," but also for machines of musical entertainment — delivering packages of pleasure. At first, this meant record playing in public, principally by means of coin-slot vending machines dispensing music.

The idea of automated coin purchasing is sometimes traced back to an ancient Egyptian device said to have dispensed a magic potion when a devotee tripped an opening with a coin of a certain weight. The prototype of the modern juke box, however, came only in November 1889 when Louis Glass, a manager of a San Francisco phonograph agency, equipped Edison phonographs with listening tubes and a coin-triggered starting device. By the summer of 1890, saloons and drug stores outfitted with such machines were offering marches, waltzes, and popular tunes recorded by small brass bands and solo in-

**Figure 5.3** Young people on the go step up for a quick hit of music at a phonograph arcade, circa 1889, probably in San Francisco. U.S. Department of Interior, National Park Service, Thomas Edison National Historical Park.

struments, along with songs sung by unnamed voices. Soon thereafter coin-operated phonographs began appearing in penny arcades. Naturally, phonograph records reproduced the popular music of the era, especially the two-minute songs performed in vaudeville and sold as sheet music for home use with piano accompaniment. For a time these coin-op record players were very profitable, earning $50 or more per week per machine. The phonograph had begun its long history as a dispenser of hits of musical entertainment.[18]

The coin-op's success was short-lived, however. The machines typically had to compete in noisy settings with devices that produced much greater volume—like the Polyphon, a coin-operated German-made music box (1893), and player pianos, whose makers adopted a common standard for prerecorded rolls in 1908. These latter gadgets freed the listener from cumbersome sound tubes while offering a broader set of listening choices. Coin-operated phonographs were subsequently fitted with sound horns and in 1906, the Automatic Machine and Tool Company of Chicago built a contraption that allowed listeners to select from among twenty-four different recordings arranged on a Ferris Wheel–like device. But phonograph parlors still declined after 1908, due principally to the lack of amplification. By then the coin-operated machines—coming to be known as "vending machines"—had largely passed on to other uses, including fortune telling and testing people's weights or capacity to lift, punch, grip, or even breathe—as in coin-operated "lung testers." Vending machines were also used for gambling and for dispensing food, cigars, cigarettes, gum, or naughty pictures. Coin-operated juke boxes would be revived with the invention of electric amplification in 1927, but in the interim this particular form of packaged public sound was not terribly popular.[19]

Far more successful was the phonograph for home use. By 1892, even Edison had begun to recognize the possibilities of sound recording for "home entertainment." Delayed by court fights and technical challenges, Edison introduced his Home Model A phonograph in December 1896, winning instant popular favor.[20]

This domestic sound-making machine was revolutionary. The phenomenon of a mass-produced durable commodity for comfort (or pleasure) in the home was itself relatively new: predecessors include the cast-iron stove, the piano, the sewing machine, and even novelty items like the stereoscope—but not a lot more than these. The appliancing of the home meant more, though, than easing domestic tasks and raising standards of living. It also had the interesting effect of *privatizing* certain types of pleasure, especially those that could be commercialized, shifting these from public places and social gatherings into private domestic spaces. The phonograph was one of the most important new home appliances insofar as it enhanced personal choice even as it created access to a larger geographic (and eventually global) culture of reproduced voice and music. All of this became first possible with mass-duplicated recordings sold to individual consumers for personal use.

Also interesting is how who-was-recording-what changed over time. While Edison continued to produce phonographs that could both record and play back, for example, other companies dropped the recording function altogether and turned the "record player" into an inducement to purchase commercial recordings. This is an early example of the "razor and razor blade" marketing that drives a great deal of merchandizing today (think of hardware vs. software). The phonograph also represented a dramatic change in entertainment based on a new business model: the mass commercial distribution of notable voices and musical ensembles from central locations to home sites. We are so used to hearing in this manner that it is easy to forget how different our world was prior to the 1890s, when music was either consumed (listened to) in public or produced (played) in private, most often by amateurs (albeit sometimes with mass-produced sheet music). The commercial recording of sound created a new kind of fusion of public and private, recreating celebrity entertainment in the parlor.

Early phonographs were expensive and unreliable, however, and the most profound impacts would have to wait until the twentieth century. Edison's cheapest model in 1893 cost $140, and even though prices would fall over time (his Standard sold for only $20 by 1898), phonographs remained beyond the means of most Americans until the 1900s and were difficult and often finicky to maintain. Consumers had to regularly adjust the brushes and governor assembly on the electric motor, and batteries had to be recharged with nitrate of soda, a messy process. For music producers, too, record duplication was no easy task. In the early 1890s, an Edison recording session required a live performance in front of as many as ten recording cylinders. The number of cylinders produced could be increased by mechanically transferring an original cut cylinder onto wax blanks using a pantograph, a device used to copy writing or engravings. Typically a singer would perform ten times into three horns; each of these 30 copies could then be reproduced about 25 times (via pantographs) to yield some 750 finished cylinders ready for sale. Attempts to create a satisfactory mold of a recorded cylinder for mass duplication failed until 1902. In the meantime, home recording (of family voices, for example) survived for a while on Edison machines but was always cumbersome. Indeed, it was difficulties such as these that reinforced the trend toward the phonograph becoming a purely "passive" playback device for delivering professionally recorded performances.[21]

Edison's rivals also embraced the idea of the phonograph as an

entertainment machine. As early as 1891, Columbia offered a catalog of recorded songs for voice and orchestra for private purchase.[22]
An equally important event was the 1895 establishment of Berliner's
American Gramophone, a company that radically changed the (physical) shape of recorded music. Berliner's disc records proved to have a
distinct advantage over Edison's and Columbia's cylinders: Berliner's
flat discs were far easier to duplicate, especially after he started making
them from inexpensive shellac instead of the more costly ebonite rubber (in 1898).[23] Flat records were also easily and cheaply pressed from
a metal matrix, reducing the costs of the final product.[24]

Home use was also encouraged by a seemingly backward development—the introduction of a nonelectric spring motor to replace the
cumbersome battery-powered motors. Eldridge Johnson built a spring
motor turntable in 1896 for Berliner, which assured a constant speed
and thus minimal distortion of the spinning disc. This was a major improvement over the cheap hand-crank Gramophones Berliner had first
made. Another innovation from the mid-1890s was the shift from the
rubber ear tube for conducting sound (something like a stethoscope) to
a trumpet-like metal horn, making it possible to enjoy a talking machine
without wearing any kind of contraption. Group listening also became
more comfortable and more convenient.[25]

All of this reinforced the shift toward passive music playback rather
than active recording—an interesting deviation from the trajectory
of the personal camera (discussed in chapter 6). By 1900, instead
of recording the voice of some beloved family member, the phonograph in tens of thousands of homes reproduced the voice or music of
some celebrity, a category itself partly created by this new technology.
Records became less a way to preserve one's personal past than a way
to keep up with novelties in the global present.

The mystique of the captured individual voice may have generated
the first wave of enthusiasm for the phonograph in the 1870s, but a second wave in the 1890s was much more about participating in a shared
culture of sound, purchased for enjoyment in a familiar and private setting. With the phonograph record, music became more like "a mono-
logue," a kind of "stockpiling of sociality" for private use, in the words
of the French economist Jacques Attali.[26]

But all this was limited by the technology of the time. Acoustic recording captured only a small spectrum of sound (168 to 2,000 cycles per second, compared to the human ear's range of 20 to 20,000), while distorting the sonorous qualities of whatever was actually captured. Records also reproduced brass and woodwind instruments with greater fidelity than strings, and midrange voices (especially tenors) sounded better than low voices or high sopranos. Recording in wax was also very sensitive to changes in volume and pitch. Loud or deep bass notes sometimes forced the stylus to jump outside the grooves in discs, ruining the wax master. And because of the need to sing or play directly into a recording horn, large ensembles were impossible. Keep in mind also that recordings from this era lasted only 2 to 4 minutes. Of course, players adapted: tubas replaced low strings, for example, and special violins with built-in diaphragms and horns were devised to increase their volume. Singers also learned to adjust their distance from the recording horn according to whether they were singing low or high notes (moving closer for the lows). One largely forgotten curiosity is that if the song was too short for the record, the band would applaud itself until the end. And artists often had to labor for hours or even days, reperforming their pieces literally dozens of times onto cylinders to create reproducible masters. Musicians willing to do this were hardly stars; many were in fact anonymous, and performers were often handed sheet music only after they arrived at the Edison studio.[27]

Constraints such as these forced limits onto the range of early packaged sound. And partially for this reason, recording studios favored genres such as (small) band music, solo instruments (or whistlers), or even comical recitations over orchestral music and grand opera. Early in the history of this technology, opera stars with any kind of reputation shunned the poor-quality recordings and the pains of studio work. The short duration of records also favored the recording of familiar portions of longer songs or instrumental pieces, as in military marches or well-known overtures. In the 1890s, Columbia signed John Philip Sousa and his U.S. Marine Band, further popularizing the tunes of the already-popular March King. Other (often unnamed) ensembles played college songs or tuneful selections from Offenbach and Verdi, along with novel songs like "Daddy Wouldn't Buy Me a Bow-Wow" and the sentimental classic "After the Ball." Pickaninny and old man Negro tunes (often performed by white singers) reflected the racist views of these minstrel-era times, but we also find oddities like the infectious "Laughing Song" recorded by the ex-slave George W. Johnson ("the Original Whistling Coon and Laughing Darkey"), who seems to have recorded it thousands of times before adequate techniques of duping were invented.[28]

One of the remarkable facts about the rise of commercially recorded music is the close link between the phonograph and the rise of new forms of popular music. Commercial songs drew from a mix of extant music forms but were also often different from the folk tunes of amateur music makers. Only shortly before the coin-op phonograph in the late 1880s, a new type of sound was being cranked out in song factories in lower Manhattan (specifically in Tin Pan Alley, on West 28th Street between Fifth and Sixth avenues). First typically performed in vaudeville and then sold as sheet music, these songs had to fit within the recordable format — meaning two or three minutes long, and so without the many verses of traditional folk songs. Novelty was also an increasing emphasis, and themes were often sentimental. Publishers and their writers were often immigrants with ethnic and social backgrounds that gave them access to a variety of traditional tunes, including African-American "coon songs" and ragtime. The object, as historian David Suisman has shown, was to create musical "hits" — meaning popular but usually ephemeral ditties. Records played a key role in creating this new music industry and the entire category of "popular music."[29]

The result was a new type of sensory experience: two- or three-minute-long "hits" of musical pleasure consumed in about the same time it took to smoke a cigarette or eat a candy bar. Records offered intense but brief hits of sound, much like bottled soda or wrapped candy. And for the first time anywhere, the voices and music of strangers could become accessible and part of everyday life, even to those who had never attended a scheduled speech or "live" performance. The captured voice actually led to a new appreciation for celebrity sound, with some loss perhaps to the stature of the lively arts and voices of ordinary people — the original, and increasingly marginalized, "producers" of music. A radical asymmetry emerges in the kinds of sounds produced and consumed (fewer people produced, while more consumed) — a contrast that would become even sharper over the course of the twentieth century.

## Selling Packaged Sound:
## The Big Three Redefine the Ear

In the generation after 1900, packaging sound (like other sensual commodities) was transformed into an industry for mass production. In the manufacture of phonographs and records, a "Big Three" emerged that anticipated the oligopolies in so many industries. These companies competed for the broad consumer middle with technological innovation, price-pointing to attract distinct income groups, a segmentation of taste communities, and the mystique of the armchair mastery of the elusive mermaid's voice.

The Big Three's dominance was laid with the formation of Edison's National Phonograph Company (1896), Columbia's absorption of American Graphophone (1896), and the consolidation of Berliner's and Eldridge Johnson's patents into the Victor Talking Machine Company (1901). Victor emerged from a complex and nasty set of lawsuits in the wake of which Johnson and Berliner won the right to manufacture Gramophones under the aegis of the Victor Company. They also created a patent pool in 1902 with Columbia, allowing the latter to produce both discs and cylinders. A final legal hurdle emerged when music composers discovered that recordings of their works were not covered by copyright, a problem solved only by legislation fixing a royalty to be paid to composers (or performers), originally 2¢ for each recording. Victor Herbert and other composers later (in 1914) joined to form the American Society of Composers, Authors and Publishers to license and collect royalties.

Each of these recording companies developed vertically integrated supply chains, erecting factories in Camden, New Jersey (Victor), Bridgeport, Connecticut (Columbia), and West Orange, New Jersey (Edison), to make cabinets, horns, and diaphragms/styluses for a wide range of phonographs. The Big Three also made their own records and distributed these—along with the players—through a national system of franchised dealers and department stores. A number of smaller companies managed to survive in the low-priced sector (the $10 Echo-phone, for example) — and in the early 1920s there were still about two hundred American record companies offering an array of music, including the short-lived African American record company, Black Swan. Still, the Big Three dominated by forcing dealers into exclusive contracts, obliging them also to conform to company-set prices and standards of service.[30]

Victor's dealers, for example, were expected to carry recordings of artists whose concert tours drew local audiences; the expectation was also that these dealers would use window displays of some featured performer (Enrico Caruso appeared in cardboard cutouts, for example) to attract customers. In 1909, the company encouraged franchisees to contact local school boards to offer record concerts to students and even to present summer park concerts, where their "talking machines" would substitute for brass bands. Victor salesmen traveled widely to find investors for main street dealerships, and Edison's National Phonograph Company did much the same for smaller cities and the hinterland.[31]

National advertising supported these stores by creating demand for the phonographs and records. As with canned foods, the makers of "canned music" (so called for the shape of the canister holding Edison's wax records) took advantage of nationally distributed magazines to introduce their novelties. Splashy full-page phonograph ads were appearing in *McClure's, Cosmopolitan*, and *Harper's Weekly* by the 1890s, and in 1901 Victor launched ads featuring its iconic "Nipper" the dog mascot, peering, head cocked, into the Victor horn in search of "His Master's Voice." Few images have been more widely circulated, and indeed, Nipper himself was flattered with countless imitators. Zonophone had a baby looking quizzically into the horn, and Talkone did the same with a parrot.[32] A 1901 ad likewise featured a small boy chopping into an Edison Phonograph with a hatchet, "looking for the band."[33]

Aggressive advertising and retail promotion familiarized consumers with the talking machines and built the Big Three into its main purveyors. But making this packaged pleasure into an everyday appliance required continuous technological advances. One key issue, as in so

**Figure 5.4** A publicity photo featuring Victor's Red Label star, tenor Enrico Caruso, beside Victor's Victrola, circa 1910. Bain News Service, Library of Congress (LC–DIG–ggbain–29837).

many subsequent media formats, concerned the recording platform: in this instance, cylinders versus flat (stackable) discs. Cylinders for a long time produced superior recordings, but they were also more expensive to manufacture. Edison found a temporary solution in 1901 with a new method for mold casting cylinders,[34] but even after his four-minute Amberol record debuted in 1908 (playing twice as long as the competition), the cylinder could not overcome inherent limitations.

Cylinders didn't have the audio volume and depth of the disc (allowed by the lateral cut), and discs were also easier to store—a nontrivial advantage in an era when listeners were starting to accumulate personal music collections. (This "stockpiling" is typical of a great deal of modern consumer culture.) Finally, in 1912, Edison gave way and started manufacturing "improved" discs (while still offering cylinders). His "unbreakable" Diamond Disc made of phenol resin was played with a long-lasting diamond stylus, superior to the steel or fiber styluses of Victor and Columbia. Edison also introduced a new cylinder, the Blue Amberol, made not from wax but from celluloid, a tough, cellulose-based plastic that could withstand hundreds of plays with relatively little loss in fidelity.[35]

Improvements such as these prefigured what consumers would later come to expect in annual changes in automotive "styles" and, even now, the continual upgrading of personal computers or smart phones. In the case of the phonograph, the "new and improved" pointed to a distinctively modern (and really unprecedented) promise: an ever more refined, convenient, and intense sensual experience. As part of this, phonograph manufacturers advertised not so much against competitors as against their own last year's model. An early Victor sales brochure bragged that there was "no comparison between the sound reproduction of even three months ago and the magnificent results produced today." An ad from 1905 boasted that the Victor "ranks with a Stradivarius." The idea was to distance the device from earlier incarnations of talking machines—used as toys or business tools. No, now it was akin to a musical instrument—and a precious antique violin at that. This made the Victor appropriate for the parlor, much like a piano as distinct from, say, a sewing machine.[36]

All of this encouraged rising expectations and a presumption of inevitable obsolescence—that the "old" would soon pale in comparison with the new. In this founding age of progressivist consumerism, personal progress required possession of the "improved"—and this meant access to the sensual delights of the packaged pleasure. But you didn't even have to wait for new models to climb the ladder of improvement. In the 1890s the big phonograph companies developed a technique copied with great success by General Motors and other car companies, rolling out a "full line" of products graduated by luxury features and price. In 1896, for example, Edison offered a range of phonographs based on motor and horn size. By 1899, he had added an entry level device (the Gem) marketed to children, with the expectation that with age and income users would graduate to the pricier models. The 1910 Amberola was priced at $200 (when the Gem's price had dropped to $5), but by 1913 the snootier set could purchase an Amberola set in classy cabinetry: the $450 Louis XVI, for example, came in Circassian Walnut with eighteenth-century-style legs.[37]

In 1900, Victor also offered a differentiated line, ranging from the Toy for a mere $3 to the upscale Monarch for the "homes of music lovers who have hitherto scorned the talking machine," costing $150. And high-end model offerings expanded greatly with the introduction of the Victrolas in 1906. The cabinetry of these upscale record players (with enclosed horns) came in a wide variety of styles (Queen Anne, William and Mary, Louis XVI, Gothic, and Chippendale) and were clearly meant to be displayed.[38] Packaged sound came in all kinds of packages befitting different sized pocketbooks.

## Edison Records: Meeting the Mass Market

By 1900, the clear purpose of the talking machine was not so much to capture the uniquely personal voice but rather to possess the sounds of an increasingly shared commercial and celebrity culture. Tastes differed of course by class, region, age, and ethnicity — and this had long predated recording. But sound recording expanded, blended, and to a certain extent standardized those taste cultures. It also accelerated the trend toward privileging the voices of celebrity. This is a complex topic that we can hardly do justice to here. But it is nonetheless worthwhile to explore some elements of this phenomenon as seen in the marketing of records. Edison and Victor (with Columbia in between) adopted very different attitudes and strategies toward recording and marketing records. Let's begin again with Edison.

Historians have long noted that Edison maintained a strong personal influence over his phonographs and catalog of records. Clearly he favored his machines, their innovations, and their reliability over his music. Edison touted his technology: "the clearest, strongest records, the most durable reproducing point and the correctly shaped horn." His phonograph, ads claimed, was "a scientific instrument [not a musical instrument, as claimed by Victor] made with great care in a laboratory which knows how to make every part right. . . . Thus it has become the greatest amusement maker ever produced." Hardware came first, and, for a time, it was seemingly outdated hardware at that. He held on to the cylinder exclusively until 1912, even as Columbia had started offering discs in 1902 for its high-end market (while selling cylinders to the popular taste until 1909).[39] Edison held out even when he had to sell his cylinders for 35¢, while Victor sold opera discs for a dollar.[40]

Edison imposed his own musical taste on his catalog and tended to favor rural, small-town preferences (where most of his franchisees were located). A 1908 ad pictured a couple celebrating their fiftieth wedding anniversary listening to an Edison (see figure 5.5); the caption read: "Some music never grows old, particularly if it recalls pleasant memories." In private, Edison insisted that the "public is very primitive in its tastes," a view certainly reflected in his preference for "coon songs." On occasion he went upscale, as in 1906 with his Grand Opera Series, but this accommodation to the "carriage trade" was always half-hearted.[41] Edison famously wrote to band leader John Philip Sousa: "All the world wants music; but it does not want Debussy; nor does it want complicated operatic arias." In 1909, Edison signed Victor Herbert to select music for recording, because this popular composer and band leader "has made the kind of good music that is likeable." He rejected world famous Sergei Rachmaninoff because of his "bravura style of pianism," which he denigrated as "pounding." He also insisted on consonant harmony and rejected "obsequious accompaniments," possibly because he was hard of hearing.[42]

Edison may well be the greatest inventor ever to have lived, but there is a short-sightedness in his approach to sound that is curious in retrospect. Perhaps most remarkably, he failed to see the point of celebrity recording. In a 1907 ad, the company described its April list of twenty-four new records simply as "the choicest bits of vocal and instrumental music recently produced together with a sprinkling of things not new but good . . . made by the best procurable talent." Edison insisted

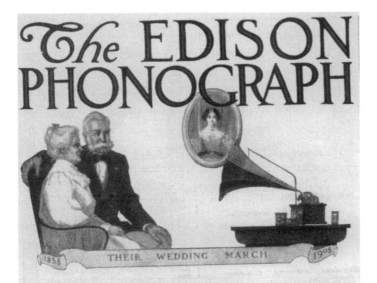

# The EDISON PHONOGRAPH

THEIR WEDDING MARCH

1858     1908

SOME music never grows old, particularly if it recalls pleasant memories. The Edison Phonograph can reproduce for you the marches, ballads and airs that stirred you in the old days, just as well as it can sing the song that is the current hit in the metropolis, doing it with a clearness, a fidelity and a beauty and volume of sound that is not to be found in any similar instrument.

The Edison Phonograph is all things to all men at all times. Simply by changing a Record it may be a brass band at one moment and a violin virtuoso the next, a singer of ragtime or of grand opera, a funny vaudeville team or a quartette singing a sentimental ballad.

If you haven't heard the Phonograph lately, you'll be surprised at the wonderful improvement in the new model Edison with the big horn. Ask your dealer to show it to you or send to us for booklet describing it.

*Thomas A. Edison*

**Figure 5.5** A typical Edison phonograph ad touting Edison's technological superiority and range of cylinders, but also appealing to nostalgia in how his machine can bring back a long lost youth. *Collier's*, March 28, 1908, 4.

in 1912 that his customers "care nothing for the reputation of the artist, singer, or instrumentalist, except in a few rare instances where the
person has established a unique and isolated position." He believed
that most of the vaunted stars recruited by Columbia and Victor were
famous for their acting, not singing. For him it was all about the tunes
and sound quality, not the "personality" or associated fame.[43]

But Edison's musical packages were hardly backward-looking. His
studio understood the power of the contemporaneous, boasting at one
point (in 1903) that it introduced twenty-five new records every month.
Like other modern packagers, Edison opined that "successful men are
believers in advertising." Edison ads from 1905 appealed especially to
the up-to-date listener, the person who had heard a new tune at the
theatre and wanted to play it at home until they had "mastered both
music and words," allowing the "fleeting pleasure" to be "made permanent." Edison understood the ethos of modern popular music and
its listeners; his professed job was to capture the ever-moving conveyer
belt of novelty, "mastering" it before it passed by.[44]

Edison also recognized the power of music to evoke and manage
emotion and mood. In 1914 in his monthly magazine to dealers, he observed that "music is more essential than literature, for the very simple
reason that music is capable of releasing in practically every human
mind, enlightened and ennobling thoughts that literature evokes in
only the most erudite minds." Next to religion, music was "the mind's
greatest solace and also its greatest inspiration." In an advertising booklet from 1921 his company offered readers a "Mood Change Chart,"
which claimed that musical sound could be broken into 13 different
categories—evoking feelings of tenderness, solemnity and joy, for example. Aided by W. V. Bingham, director of applied psychology at the
Carnegie Institute of Technology, he selected 135 records from the Diamond Disc Collection and arranged these according to the psychological effects they supposedly had on the listener. But Edison never
appreciated the appeal of the celebrity musician. What he did understand was the value of packaged sound for mood control—anticipating
Muzak in this respect—and the modern need to "keep up" with changing fashions.[45]

## Victor Records: Leading with Class

By contrast, the Victor Talking Machine Company had no such doubts about the appeal of celebrity and of "class." Victor's success, however, was especially in its "software," its records, and here it emphatically staked out the prestige trade with celebrity recordings. This began in 1902, when Fred Gaisberg of British Gramophone (Victor's affiliate in England) signed Enrico Caruso of the La Scala Opera House in Milan. Caruso became the signature artist for Victor's upscale Red Seal label, a star a decade before the star system hit the movies. Nellie Melba, Emma Calvé, and Antonio Scotti joined Caruso in 1904, while the rising tenor John McCormack and virtuoso violinist Fritz Kreisler were signed in 1910, all with much hoopla. Names such as these were publicized in ads featuring these musical greats all gathered together on a single stage, listening to themselves on Victor machines. Victor also followed a well-established tradition in winning the endorsement of royalty (England's George V) as well as famous composers (Edward Elgar and John Philip Sousa). Victor even offered dealers instructions on how to pronounce the names of artists and composers to win the confidence of discerning consumers.[46] David Suisman describes the simplicity of Victor's formula: "To sell machines . . . Victor pushed records, and to sell records, it promoted high culture."[47]

But this was more than snob appeal. Victor understood that these stars brought voice quality and prestige, but the company insisted also that their voices evoked the singers' lives and unique selves. Victor's magazine ads frequently showed these celebrities in costume, sometimes as miniatures, almost dolls, emerging from Victor sound horns or standing on Victrolas as if they were the playthings of the listener. Caruso records brought "not only his art but his personality." Record buyers (like the sea witch) owned the "throat of Patti, Melba, Calvé and the great dramatic tenors Caruso and Tamagno." In 1913 Victor bragged that it *was* Caruso.[48]

As central to Victor's brand was its appeal to status and class. "Years of patient experiment—with various woods, with different proportions, with numerous vibratory surfaces" yielded, according to one Victrola ad from 1911, "that same indescribable 'something' which makes the Stradivarius supreme among violins, gives to the Victor Victrola such a wonderfully sweet, clear and mellow tone as was never known before." While Edison had hyped the heroic dedication and scientific integrity of its genius inventor, Victor appealed to the status-conscious affluent by promising heritage and craftsmanship.[49]

Crucial to Victor's image was its success in adapting the phonograph to the late-Victorian bourgeois parlor. Long the site of the piano and fine furniture harking back to eighteenth-century aristocratic fashions, the parlor was certainly not a place for a modern machine. Company leader Eldridge Johnson had to convince Victorian tastemakers that his contraption fit in this "traditional" and "timeless" setting. Key here was the introduction of the Victrola in 1906, with its mechanical works hidden in a four-foot-high cabinet of piano-finish mahogany. Sound was projected out of the front of the cabinet, and a wooden lid hid the turntable. A Victor trade catalog of 1908 suggested that the Victrola was suitable for the "drawing room" in that it "conveys no hint of its purpose" except for the crank for the spring motor. Victor extolled the Victrola as an "artistic cabinet" and fine piece of furniture, an exquisite example of Victorian sensibility insofar as the industrial origins of comfort were hidden in an increasingly industrialized world. Victor regularly pressured dealers into remodeling their stores to reflect the elegance of the contemporary bourgeois parlor (complete with listening rooms with leather chairs and soft lighting). One series of ads even showed the Victrola in the White House. And once established as legitimate for the parlor, the Victrola began to replace the piano (and player piano) as the fashionable instrument of genteel domestic gatherings.[50]

At the same time, Victor did more than exploit the sentiments and prejudices of bourgeois domesticity. Johnson's staff took on the role of music education with the publication of the *Victor Book of the Opera*, which included plot summaries as well as promotions of releases. The company also published *Music Appreciation with the Victrola for Children for Elementary Grades* and a *Music Manual for Rural Schools with the Victrola*, outlining its mission to have children "hear only that which is good, refined and cultured." Victor in 1911 actually established an Education Department to help teachers and parents raise money to buy these still quite expensive machines. Mechanized reproduction of music was held to have pedagogic virtues, as when a 1912 Victor ad insisted that Tchaikovsky's Fourth Symphony must be listened to several times to enjoy and understand, which of course meant purchase of the company's Red Seal records.[51]

Victor also catered to less-affluent tastes by segmenting its product line. The Black Label provided an alternative of popular music, drawing from the vast reservoir of tunes turned out by Tin Pan Alley. The 1903 catalog included cartoonlike drawings of bands and minstrels, promising "dear old Southern melodies sung and played upon the Victor." Later featured numbers included "Carry Me Back to Old Virginny" and "Dixie," as well as "Little Alabama Coon" and "Old Black Joe." Victor also published comic songs with stereotypical ethnic themes: Dan Quinn's "The Mick Threw the Brick" and "When Reuben Comes to Town," for example. In 1911, Victor signed the legendary vaudevillian George M. Cohan, bragging that they had bagged the "Shakespeare of musical comedy" and promising consumers "all those little Cohan mannerisms" as well as great tunes.[52]

Victor's mastery of the market was rooted in its trademark domination of packaged music, and that required expanding its audience beyond the carriage trade. In 1909, Johnson noted that Victor's advantage came from its "large capital, large manufacturing plants and most of all a well-chosen and well organized army of experts," even more than its patented technology. Victor also had a very effective marketing department and one of the most widely recognized trademarks.[53] So whereas Edison had essentially made a trademark of himself—his signature and picture adorn many of his early products—Victor by 1913 could fairly brag that it had the "best trade mark in the world" in "His Master's Voice." The image was compact, charming, and instantly recognizable. Ads simply advised to "Look for the dog" when purchasing a talking machine. In pushing its dealers to display a cutout of the mascot, Victor confidently affirmed that "the people think about the things they are constantly told to think about."[54]

But success was more than about retail domination and effective advertising. Victor understood that they were in the razor and razor blade business, selling phonographs to get people to buy records. The real "packaged pleasure" was the music, and the machine was really just a delivery device. This was made plain when Victor introduced its Victor IV model in 1911. Priced at $15, it was hardly a profitable item for dealers. But the company insisted that this entry-level machine could serve as a wedge or commercial conduit, opening "homes to Victor Music." Dealers were reminded that wage earners, too, needed music, and that they were "just as human as the capitalist and the people of higher incomes." This is easy to discount as self-serving; it certainly wasn't novel. Indeed, Victor was just following down a path blazed by Ford with his mass-produced Model-T of 1907. But Johnson was doing more than selling phonographs to the masses; he realized that marketing records to owners of Victor IV machines could be "just as profitable . . . as to those who own higher priced instruments."[55]

## The Contradictions of Captured Sounds

The range of choice in phonograph records grew at a remarkable rate in the early decades of the twentieth century. By 1923, a Columbia catalog listed 416 pages' worth of records in small print. Dixieland and other bands, banjo players, comics, yodelers, whistlers, minstrels, and Hawaiian-style singers filled the pages, along with titles by the Chicago and Columbia Symphony orchestras. Columbia offered sentimental ethnic tunes for immigrants and a wide variety of college songs for the varsity set. Appearing on these pages also were recordings by celebrities of all stripes, from vaudevillians like Al Jolson to the cellist Pablo Cassals. Though much recording was conventional and stereotypical, it did introduce white Americans and then the world to African American jazz, ragtime, and blues.[56]

But the phonograph record also set new expectations of aural "perfection," as when an ad writer for Columbia in 1916 claimed that the company's records reproduced sound "as faithful to life as the reflection trembling on the surface of a sunlit woodland stream."[57] Ads also promised an extension of human faculties and the transcendence of impermanence. An Edison ad from 1899 celebrated the phonograph as making "permanent the otherwise fleeting pleasure," offering older customers records that could bring back to life those "marches, ballads and arias that stirred you in the old days."[58]

But capturing sound went beyond nostalgia. The claim that "Only Victor owners can command this performance" appealed to a popular longing for possession anywhere, at any time, and at reasonable prices. So while Caruso might command a "princely salary" every time he sang in public, the owner of a Victor phonograph could hear Caruso "whenever, wherever, and as often as you like," with every playing of the record being a command performance. The Victor Talking Machine was like a genie's magic lamp that, once the tone arm was set onto a disc, caused opera stars and minstrels to march forth from the trade-marked Victor horn. Victor brought Milan's La Scala opera house and its "magic atmosphere" of "indulgence, of lawless emotion" to the solitary Victorian listener sitting in his or her parlor in Dubuque, Iowa. Nipper delivered a "Vaudeville pick-me-up" to rouse the tired farmer on a homestead in Montana. Even for New York City swells, the old prestige of attending a performance was to be replaced by the new mystique of the private airing of the genie in the groove. The "atmosphere" of the Met was to be "brought into your home" to be enjoyed "at a moment's notice, immediately dispelling daily woes." Victor saw itself as spreading music from the realm of a "few thousand people in a darkened hall to "lonely pioneers, travelers in the far places," even "music lovers huddled around the fire in snowbound farm houses." This crusading message was clearly self-serving, but it also expressed an ethos that really only begins with the packaged pleasures revolution—that curious combination of consumerist egalitarianism and privatized, personal satisfaction.[59]

We have seen how the record industry created new and attractive ways of listening, but manufacturers also quickly learned how to stave off satiety as repetition turned novelty into a bore. In a 1918 pamphlet, titled "How to Get the Most Out of your Victrola," Victor instructed consumers on how to go beyond the novelty of the popular tune or operatic piece. Listening henceforth had to be an active process; consumers had to learn the story behind the characters and their arias. But even Victor realized that to "become saturated with anything is to lose the fine edge of enjoyment." The more lasting (and profitable) solution to the problem of repetition was to collect an ever-increasing variety of recorded voices. Here Edison's staff took the lead, suggesting that consumers invite friends and families to record parties featuring a wide range of sounds: "you can make up your own programmes" to appease all the family and to stir a variety of moods. The phonograph-playing host could be like a vaudeville impresario, offering popular variety as distinct from the "refinement" promoted by Victor's taste makers. Any mood, however ephemeral, could be evoked and then discarded simply by changing the record as stressed in Edison ads.[60] Of course the phonograph may also have shortened the attention spans of some listeners—who could now access the "hit" song without any need to wait for or endure a long concert.

At the same time, the phonograph was supposed to uphold traditional Victorian family ideals, even as it transformed the home into a distinct personal site of pleasure. With the phonograph, the home could become more than an escape from job, market, and public strife; it could now become a haven of family togetherness. The home also became the narrow point of a funnel drawing in a global celebrity culture for personal consumption. This combination of the private and the global would be consolidated and extended with the coming of radio, television, cable entertainment, and the Internet. But it began with the recorded voice of the timely yet in-time, comfortably familiar celebrity heard in millions of parlors.[61]

The appeal of novelty and "star power" might have clashed with the traditional Victorian ideals of the comfort and privacy of the home, especially insofar as the late Victorian home was expected to be a quiet retreat from the ever-changing and boisterous street.[62] But the phonograph was supposed to transcend this gap. While continuing to appeal to the new and famous "out there," phonograph companies sold these new machines as a way to rescue Victorian domestic ideals from the "threat" of disorderly public space.

Edison's ads repeatedly made this claim: "Make home a competitor of downtown, the club, the café, the theatre and the concert hall. No one thing will furnish so much amusement for so many people, so many times, and in so many ways as the Edison Phonograph." Home

**Figure 5.6** A fantasy of possessing the glamour, personality, and voice of Victor's stable of world-famous opera singers in the privacy and comfort of your parlor is fully expressed in this ad. Note the iconic image of "His Master's Voice" in the corner. *Collier's*, April 11, 1913, back cover; *Saturday Evening Post*, September 19, 1914, back cover.

music was to be a soothing balm to the stresses of market or public life. And while public entertainment increasingly divided the family, the phonograph could make it whole again. Appealing to the breadwinner, recorded music was supposed to be "as much a part of your home as the wife and children," linking "old and young together."[63]

Particularly important in creating this domestic aura was linking records to growing children. Edison in 1908 exhorted: "Don't let the young folks get into the habit of seeking amusement outside of their own homes." The phonograph was to be more than an alternative to the amusements of the street; it was also to be a companion to the child, especially as the size of families decreased and parents felt ever-greater needs to distract or "entertain" their kids. Edison insisted that a record player was "the best play fellow a child can have." Indeed, the phonograph was supposed to be in "every home where there is even one child."[64] The Edison Company recognized that children were on the cutting edge of an advancing consumer culture, advising parents with regard to their offspring: "Take them with you when you buy" records and let them display their knowledge of the latest thing. In 1912, Edison ad makers admonished parents to record the sayings of their children and instructed kids directly to "Record your progress as you sing or play." Phonographs would build family ties even as they gave increasingly indulgent parents an opportunity to encourage their children to express themselves.[65]

Yet the phonograph, like its successors in the form of radio and TV, while first touted as an instrument of family togetherness, often split the family apart. Lowered costs also allowed the proliferation of devices, and it became increasingly common for households to have more than one record player. And portability fostered newly specialized "niche" uses. As early as 1913, Victor promoted small Victoria models for summer listening on the porch, at the dance, even on top of a boat and at campsites. By becoming portable and escaping the confines of the Victorian parlor, the record player began the long trend away from family-shared entertainment and helped to create "generation gaps" in musical tastes.[66]

## The Captive Voice

By containing the human voice and music, the phonograph was a triumph of technology over ephemeral nature, lifting sound from its customary place and time. For the first time since the evolution of speech, the human voice could be separated from the organs that had always generated it — to the astonishment even of dogs and parrots. The human experience of music and speech was transformed in the process, allowing us to concentrate and manipulate sounds in ways not previously possible, to hear things in new ways and places, and to substitute the voices of professionals for the voices of amateurs.

Records also pinpointed sound in ways that we tend to forget. Of course, the labor of selecting and placing a cylinder or disc on the machine and adjusting the stylus created its own expectations, but the brevity of sound, trimmed to two to four minutes, introduced a pace of hearing that was new and unaccustomed. For Victor's Johnson, this progressive confinement and concentration of sound was a virtue, given that "future generations will be able to condense within the space of twenty minutes a tone picture of a single lifetime: five minutes of a child's prattle, five of the boy's exultations, five of the man's reflections, and five from the feeble utterances of the death-bed." A curious idea, perhaps, but reflective of this push to package sentiment, here with the added assumption that a life could be encapsulated in some summary quintessence. This was joined with the romantic idea of preserving a unique personality in the voice and moment of speech. Between 1889 and 1912, phonograph recordings of famous voices ranged from the 1904 campaign speeches of W. J. Bryan and W. H. Taft to the utterances of Queen Victoria or chestnuts from the likes of Florence Nightingale, Mark Twain, and Robert Browning. This notion of representing a life in an aural essence — a few choice words, spoken with characteristic timbre, inflection, and style — was not invented by records: signature vocal expressions had become shorthands for celebrity in music hall and vaudeville a generation before and "trademark" sayings or tunes marked the identity of radio and film performers a little later (think of Jimmy Durante's "Inka Dinka Doo" or Bob Hope's "Thanks for the Memories").[67]

The phonograph record also wildly advanced celebrity culture—and opposition to it.[68] Enrico Caruso in 1904 became the world's first artist to sell a million records, by which time critics had started agonizing over how the phonograph might erode the musical skill of amateurs. Even Sousa, though professionally associated with phonograph recording from the beginning, feared in a 1906 article that the phonograph was a substitute for "human skill, intelligence and soul." And there seems to have been some confirmation of this claim: sheet music dropped in price from 40¢ to 10¢ from 1902 to 1916, partly because of the decline of parlor soirees of amateur piano and voice, while phonograph sales rose from $27 million to $158 million from 1914 to 1919. And even though the phonograph gradually lost its association with toys or music boxes, some intellectuals and musicians continued to treat its use as a sign of laziness. Gradually, however, the skill and connoisseurship involved in record collecting and informed listening became something of a substitute for musical training and performance. The accumulation of records became part of a broader culture of collecting (as of photos), a central element of an emerging consumer culture. Owning a well-considered collection exhibited wealth and taste, of course, but it also confirmed membership in a national and eventually global community of the cultured or initiated.[69] The professional recorded, but the cultivated amateur could exhibit good taste by an appropriate collection, well-chosen playback gear, and "music appreciation"—an expression whose popularity skyrockets from about 1910.

Finally, the record severed listening from performance in both place and time, disembodying packaged sound even as it made it portable and repeatable. Freed from the traditional settings of live music (church, concert hall, café, or domestic chamber or drawing room), the phonographic record distanced music from its ceremonious contexts in processions, liturgical rites, celebrations of war and victories, public dances, or decorous social gatherings. Before recordings, music occurred at special moments in time during which it shaped the emotions of convened groups. By the early nineteenth century, however, live music (like visual art) was more often deliberately "framed," performed in settings apart from churches, for example, and in other non-ritual contexts. Musical performances came to be programmed into "musical time periods"—as in scheduled concerts (typically for pay).

A far greater change occurred with the arrival of the phonograph. Recorded music privatized and decontextualized what had often been a social event—and church music or sailors' songs could now be heard while dressing in the morning, while Beethoven was transformed into "wall paper." At its simplest level music was *heard*, but performers were no longer *seen*. Sounds of all sorts could thus be integrated into everyday life as a kind of aural background—but more often to enhance a mood or decorate some setting than as a focused object of attention. Claude Debussy in 1913 fretted that recorded music could be bought as easily as "one can buy a glass of beer." And with no hint of irony, the first issue of *Gramophone* (1923) called for listening to recorded music while shaving.[70]

This disembodiment of sound had other effects, and not all of them disconcerting. By some estimates, music based on oral tradition (such as that in the Middle Ages) had previously seldom been performed for more than one or two generations. By contrast, with the coming of records not only were the "classics" preserved, but formerly ephemeral "hits" could now survive as "oldies." So the music of one generation's youth could be replayed years later to evoke memories in some (often quite narrow) age group of people. Bach, Beethoven, and Brahms of course survived well beyond their social and cultural settings, taking on new meanings in the recordable world. Memories of falling in love or just being free and having fun also came to be linked to tunes that could be recalled or replayed again and again throughout life, thanks to the recording.[71]

Records in this sense fulfilled an age-old longing for containing, controlling, and ultimately intensifying sound, perhaps the most ephemeral of sensations. But their prominent place in modern consumer society appeared only after several false starts. The phonograph began primarily as a tool of corporate business enterprise, and it was not until the 1890s that it was improved for use in an emerging consumer market, as an entertainment-delivery device. And with its conversion into a domestic appliance, the phonograph helped transform the home into a privatized site of leisure and into a target for a vast, rapidly changing, but centralized mass-culture industry. The phonograph became one of many commodities linking corporate production and privatized consumption, a dyad created by hollowing out much of the social and customary life in between. Recorded sound was disembodied, divorced from ritual and sociable time and space, and thereby became immediately accessible and eventually portable. Recordings gave us aural "hits" disassociated from sight and other sensations once bound to sounds, even as these hits became part of the warp and woof of new forms of daily life. Packaged music makers promised to preserve and even enhance tradition with the "classics" played on the Victrola in the parlor. Its promoters insisted that the phonograph would revitalize the family around a common enjoyment of music and speech—and we do find recorded sound creating new kinds of conviviality. But talking machines very often privatized listening, and their producers homogenized while simultaneously segmenting tastes.

Even as it disseminated "high culture" (at least to those who could afford it), the infant record industry vastly expanded the range and impact of the pop tune and the celebrity to the detriment of community singing and other, more localized or home-grown, musical traditions. The packaging of sound did not have the dramatic bodily consequences of the syringe of heroin or the rolled tube of tobacco or even the calorie-loaded candy bar, but it did dramatically change how we hear and think about sound in ways we too often ignore.

# 6

# Packaging Sight

*Projections, Snapshots, and Motion Pictures*

We are primarily a culture of sight. This is obvious in a stroll through any local gambling establishment, like the Hollywood Casino near Harrisburg, Pennsylvania. Nothing extraordinary really: the standard flashing lights of slot machines into which hundreds of emotionless faces stare. Even the clink of coins has disappeared as bets are made and payoffs are paid electronically, on paper tickets. Only a curious whirling sound peppered by subtle electronic bells breaks the eerie silence of the players. On an upper level is a grandstand with a full view of an oval racetrack, totally enclosed from the smells and sounds of horses and the race, but with immediate access to multiple screens of race forms, race results, and images of tracks across the country—a perfect example of the tyranny of sight in the modern world. This "new perceptual field" emerged, according to Donald Lowe, in seventeenth-century Europe, followed by the diminution of smells and sounds in the emerging urban public spaces.[1]

It is hard to understand modernity without understanding a kind of obsessive exploitation of visuality. Technologies for extending the visual for elites were developed in the seventeenth century—think telescope and microscope; later these technologies gave the less than wealthy similar visual enlargements. Eyeglasses for ship captains were common by the eighteenth century, and binoculars were used for hunting and exploration shortly thereafter, but all this also suggests entirely new ways for humans to "see" the world. Sight no longer had to be ephemeral or limited to the individual body's normal range of experience, as new technologies extended vision far beyond what even the seasoned traveler could encounter. A new age of intensified and extended imagery liberated people from the visual constraints of the past and brought much of the world—or at least its appearance—into immediate experience in ways never before even imagined.

Murals and framed paintings and other decorations of the home, church, and palace brought visual rarities into everyday experience, but we should not forget how uncommon such images must have been prior to ubiquitous photography and television. Village festivals may have featured reds and blues and exotic shapes and designs—relatively rare in nature—but then so too were festivals rare and limited by local artistic skill. As in so much else about sensibility, people strove to go further, to extend the gaze and to make it mobile, shareable, and permanent. This all culminated in our era of the packaged pleasure in the key innovations of the personal celluloid-film camera and the publicly exhibited motion picture. The historical backdrop to these innovations tells us much about human desire and its technological constraints until modern times.

## Seeing More, Longer, and Whenever

When we think of technology, we mostly imagine machines that replaced manual labor and increased the output of goods. But long before industrialization, people used tools to preserve and extend the power of the human sensorium. And in the packaged pleasure revolution, as we have seen, this is accelerated through a process of intensifying and democratizing rare, but natural, psychotropic experiences, or novel tastes, smells, and sounds. Similar technologies extended our power of vision, as technological advances blended and blurred pursuit of science with pursuit of pleasure. Fascination with the mechanics of the human eye and human vision, and how those might be extended and

**Figure 6.1** A seventeenth-century image of a camera obscura, revealing how the likeness in reverse of the exterior of a "dark room" appears on its interior through a pinhole. From the Rosenwald Collection of the Rare Book Division of the Library of Congress.

artificially produced and preserved, spurred scientific investigation in the sixteenth and seventeenth centuries. But science and entertainment were never radically separate in this realm.

Perhaps the best example is the camera obscura. Oddly enough, an upside down image of an exterior can be projected on the interior wall of a darkened room (thus camera *obscura*) from the light of a pinhole on the opposite wall. Mo Ti of China in the fifth century BCE seems to have been the first to discover this curiosity, though it was not mentioned again in China until the ninth century CE. The camera obscura was known to the Arabs in the tenth century, and in the thirteenth century Roger Bacon in England used it for observing eclipses of the sun.[2] However, only with Giambattista della Porta (1543–1615) was this projection of a scene into an interior chamber popularized and turned into an object of entertainment. Della Porta managed to concentrate and turn upright the image streaming through the pinhole with an intervening lens. Observed in the private confines of a darkened room and cut off from the exterior world, the image must have been eerily enchanting. Thus was created the world's first artificial, "disembodied" image. Of course people had been painting and sculpting for tens of thousands of years, and mirrors had captured images, but here was the first direct projection of an image, and a perfect reproduction, through an instrument of human contrivance. And with seemingly perfect fidelity, opening up all possibilities of manipulation. It was the distant predecessor of today's gaze at the movie, TV, and computer screen.[3]

The camera obscura was the trunk of two distinct branches of visual technology. First, in the seventeenth and eighteenth centuries, the dark room camera obscura was reduced to a smaller, portable box (even eliminating the box in later mirror-based versions, like the camera lucida) that could be used to project an image onto paper or polished stone. That image could then be traced and eventually engraved (and thereby copied). With suitable adaptations, these images could be reduced or enlarged at will. This ability to "capture" and trace the image of whatever the camera lucida was pointed at set the stage for the photographic camera, which eliminated tracing in favor of a chemical fixing of the image. This modern version of the camera became the ideal personal tool to select and fix an image.[4]

A second branch from the camera obscura was the magic lantern, invented in the 1650s. This device retained the projected image, darkened room, and public performance of the camera obscura but eliminated one fundamental shortcoming of the earlier device. Because it required intense daylight and only reflected the exterior to the room, the camera obscura had limited entertainment value. By contrast, the magic lantern involved an artificial lamp (originally a candle or oil lamp) set behind a lens that would project a concentrated beam of light through a painted glass slide. A second lens would reverse and project the image onto a wall, allowing a near-infinite variety of artificial images, just by changing the slide—and all without natural light. First introduced by the Dutch mathematician Christiaan Huygens, the magic lantern also allowed images to be viewed by large crowds, and at will.

Promoted by the Danish mathematician Thomas Walgenstein in the 1660s, magic lantern shows became part of an emerging culture of exhibition in the hands of savants and showmen. These shows appealed to seekers of enlightenment with scenes of castles, sea ports, ships on fire, or the Saint Peter's Basilica in Rome, accompanied by lectures by some expert (anticipating entertainment as a substitute for travel). While extolling the modern and scientific character of their magic lanterns, the showmen also used their projected slides to tell fairy tales and perform gags, sometimes using titillating images of the risqué and even scatological. Though first restricted to gatherings of the well-born, a privatized variant, the *boites d'optique* (optical box, or "peepshow") offered a cheaper and more vulgar entertainment. Strapped to the backs of Savoyards, itinerant showmen in eighteenth-century France, many from Savoy, these peepshow boxes offered the common folk images of fantasy and comedy.[5]

The magic lantern in Europe began to decline about 1820, and the Savoyards had disappeared by 1850. As was often the case with adult pastimes and novelties, the magic lantern was eventually transformed into a children's toy, which "improving" Victorian parents gave to their curious or high-achieving children. John Pepper's *Scientific Amusements for Young People* (1868) offered, along with experiments in chemistry, electricity, and magnetism, detailed instructions on how to make slides and how to present magic lantern shows to appreciative friends and family.[6]

But the thrill of the magic lantern survived with adults as well, especially in Britain and the United States. In 1827, Philip Carpenter, an English optician, standardized lanterns for home and exhibition hall use. Shows gained prestige when Michael Faraday, father of modern electricity, used magic lantern slides to explain electricity in public lectures. The slide projector became more sophisticated in the 1850s, when projections of superimposed images from side-by-side lanterns could show the change of day into night. And with the development of photography on glass in the late 1840s, realistic images began to replace artistic or cartoonish drawings on magic lantern slides. By the 1880s, powerful arc lamps of oxyhydrogen or oxyether replaced smoky oil lamps.[7]

Often shown in churches, lodges, and schools, as well as public auditoria, exhibitors brought a wide variety of visual entertainment to

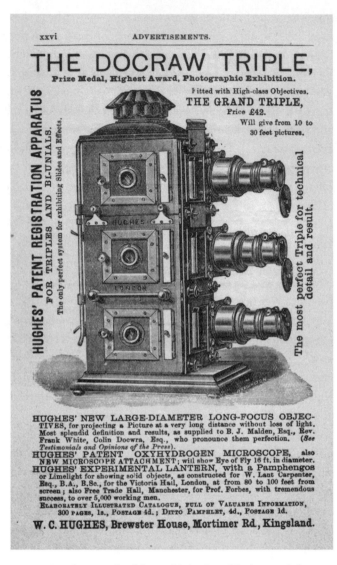

**Figure 6.2** An example of the sophistication of the late magic lantern, capable of showing multiple slides at once. T. C. Hepworth, *Book of the Lantern* (New York, 1889), xvi.

the most isolated American towns. C. T. Milligan's *Illustrated Catalogue of Magic Lantern Apparatus* (1882) offered an extraordinary range of images for sale: slides of birds, geological eras, and the planets appealed to the science minded; there were also color slides of the famous ruins of Rome and Egypt, views of the American sublime (Yellowstone, Santa Monica Beach, and Niagara), as well as scenes from the 1876 Centennial Exhibition. Slide sets illustrated stories, including the life of Napoleon, the assassination of Lincoln, and, of course, the Bible. Milligan's catalog also offered dissolving slides — one slide faded while another appeared slowly — and "slip slides" that were superimposed over one another to create an illusion of motion, such as a boat moving across a stormy sea. Many of these were "comical," as when a bull would be shown tossing a fisherman or a mother was shown beating her child. The range of lanterns by price and features was truly amazing. In 1882, the most advanced using oxyhydrogen technology for the lighting cost $350, an amazing sum.[8]

The magic lantern packaged and intensified sight. It was simultaneously science and entertainment, offering a kind of "sight travel" without the time, bother, and bodily discomforts of real-world travel. Magic lantern shows offered iconic images that the viewer mentally could tick off a must-see list. The grand tour of Rome, the Pyramids, and the Parthenon could all be seen without leaving home. Slide shows may have lacked the realism of motion pictures, but they often had the advantage of color and, in a commercial setting, were invariably accompanied by dramatic lecturers.

The magic lantern accustomed crowds to expect visual immersion in realistic artifice. The next logical step was to create an image that simulated depth, inviting the viewer to step into it. The panorama, a circular or semicircular painting, was designed to create this illusion. The idea of a massive, dramatic quasi-narrative painting seems to have originated in eighteenth-century Italy. These superpaintings of natural wonders, disasters, and battles toured nineteenth-century America and Britain. The word "panorama" itself, meaning "viewing all," was coined in the late eighteenth century, and was first used in reference to a curved or circular painted vista seen from an elevated point. Art historian Stephan Oettermann notes that this was part of emerging interest in broadening horizons during the eighteenth century: "Church spires no longer directed the gaze of the faithful heavenward; instead of looking up, human beings, themselves became godlike, now looked down from towers that served their need to see." Still later, people would build towers (like Eiffel's 1889 version in Paris) simply to satisfy their craving to "experience the horizon." And when towers were not enough, balloonists took off in lighter-than-air contraptions (beginning in 1783) or climbed mountains, bringing with them the traveler's spyglass or binoculars. The panorama was the artist's virtualist response to the discovery of the horizon, creating a sheltered and accessible replica as a visual package, "an apparatus for teaching people how to see" and thus anticipated the projected motion picture. The panorama made the viewer aware of how limited ordinary experience was and what was possible. This again, according to Oettermann, was part of the new bourgeois feeling of liberty, transcending frontiers. Pierre Prévost's pioneering panorama offered ordinary Parisians a simulated view of the Louvre from the roof of the neighboring royal palace, the Tuileries. Here was a perspective that hitherto had been available (in reality) only to the royals and their court.[9]

The panorama, first introduced in London by Robert Barker in the 1790s, went beyond the ordinary painting or even the *trompe l'oeil*. It completed the realistic illusion by eliminating the picture "frame." Patrons approached the panorama from a long, dimly lit corridor, climbing up stairs to a platform. Above was an umbrella-shaped roof (velum) that blocked out the upper edge of the canvas and hid the ring of skylights in the ceiling that illuminated the painting. Measuring at fifty to sixty feet high and ninety to a hundred feet in length, the panorama was large enough to look as if it were real. The lower edge of the canvas was obscured by the projecting platform, or "false terrain," between the platform and painting. Barker's panorama of the raging sea was a particularly big hit. Soon, other panoramas of Paris or Roman ruins attracted the throngs, often enhanced by the illusion of depth with foreground paintings and props. The rage spread rapidly throughout Europe and America. While the panorama declined in the 1830s as a commercial spectacle, interest was revived after the Civil War, espe-

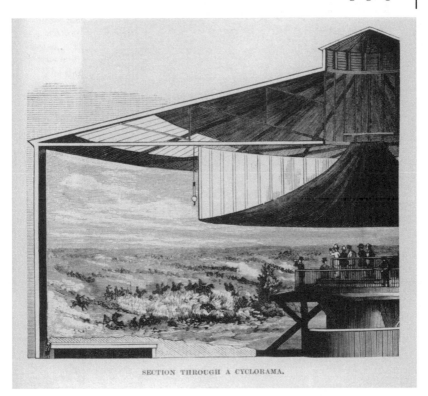

SECTION THROUGH A CYCLORAMA.

**Figure 6.3** A drawing of a panorama, or cyclorama, of a battle scene appealing to the public's desire for visual depth and realism. Notice the velum covering the top of the painting to help create the illusion of reality. Albert Hopkins, *Magic: Stage Illusions, Special Effects and Trick Photography* (New York, 1898), 357.

cially in scenes like "The Battle of Gettysburg," displayed in New York City in 1886.[10]

Popular especially in the United States were moving paintings (or "extended panoramas"): large strips of continuous landscape painting that were passed by giant, vertical, scroll-like rollers across a stage as a lecturer told stories about the changing scenery. These moving paintings were essentially travelogues, depicting the sublimity of the American wilds and frontier river systems rather than the architectural sublime of the Old World. Samuel A. Hudson began the craze in

**Figure 6.4** A drawing of a "moving painting," or panorama popular in the late 1840s in the United States, showing how the extended painting scrolled across a stage. *Scientific American*, December 16, 1848, 100.

St. Louis in 1848, with his purported "12,000 foot long painting," depicting travel on the scenic Hudson River. Soon thereafter, John Banvard, John R. Smith, and William Burr offered moving paintings of the Mississippi, Ohio, and Great Lakes, ballyhooing that their paintings were three or even four miles long. Painter-exhibitors told stories of the perils and discomfort but also the romance of traveling in an open skiff while sketching scenes along the river bank. Burr made the point well: when viewing such moving paintings "the eye, by a kind of pictorial license, glid[es] from one interesting point to another, . . . jumping over the intermediate space."[11]

However, the gigantic painting, no matter how striking and iconic, remained merely that, a painting, a fixed, albeit rolling, scene and often not of very high quality. As early as 1823, the panorama was challenged by more elaborate visual spectacles, suggesting a more dramatic kind of motion. Louis Daguerre and Charles Bouton, French veterans of theatrical set design, developed the "diorama." Although the term is used also to describe a more elaborate panorama, by 1831, Daguerre created a very special "diorama of double effect," showing on one painting two superimposed images. His first portrayed an empty church by day and a full church attending midnight Mass. The scene, painted on both the front and back of a translucent surface, could be illuminated in such a way as to reveal, in alternating fashion, different parts of the painting, according to what color of lighting was used. By gradually shifting colored light from the front to the back of the painting, images could be made to appear and disappear. A row of monks painted in light green, for example, was invisible when the diorama was lit from the front. But if a red light was shown from the rear, the monks would suddenly appear. By lighting one figure after another, the spectator would see a procession of monks pass into the church.[12]

The diorama and panorama took many other forms, creating a realistic impression of depth with scenery of imitation grass, trees, model buildings, and even live animals added to the foreground. These illusions in effect compressed time and space.[13] These visual spectacles "could be experienced repeatedly and without risk," a kind of voyeuristic spectacle that, if originally experienced with awe, could, with repetition, become banal or even boring.[14]

The magic lantern, panorama, moving painting, and diorama dazzled the eye of the novice, and with clever gadgetry could create astonishing illusions. There were limitations of course: they hardly could re-create the sensation of a dancing child or a realistic chase of robbers by cops. Long before the motion picture, audiences longed for—and sometimes managed to capture glimmers of—the illusion of realistic movement. As early as the sixteenth century, Parisian merchants hung *lanterne vive* outside their shops. These "living lanterns" consisted of paper cylinders covered in painted images (often devils), with candles lit in the interior to illuminate and project these images onto surrounding walls and floors. Air heated from the candle would turn the paper cylinder by means of fan blades fixed at the top, creating the illusion of dancing demons, attracting and fascinating passers-by.[15]

More sophisticated were the parlor optical miniatures of the nineteenth century, designed to amuse by simulating superimposition or motion. The thaumatrope ("magic turner"), introduced in London by

John Paris in 1825, capitalized on an interesting optical illusion—the persistence of an image slightly longer than its visual appearance, creating the illusion of superposition. A disc with drawings of a bird on one side and a cage on the other, when spun, created the illusion of seeing a bird in a cage. The same principle applied when, in the 1830s, Joseph Plateau of Belgium created the phenakistoscope, two discs divided into eight or sixteen sections, the first containing small slits and the second a figure posed in different states of motion. As the discs were spun, a viewer looking through the slits saw an illusion of an actually moving figure. Other contemporaneous versions of this include the zoetrope, consisting of a slitted hollow cylinder inside of which was placed a strip of paper, upon which appeared drawings of a person in a sequence of movements. When rotated, the zoetrope created the illusion of a person running or dancing, approximating what would soon thereafter be achieved with moving pictures.

Most of these optical miniatures were novelties—toys, really. Despite their appeal, the zoetrope and its cousins could not tell stories, and their motion was brief and of limited visual interest or complexity. Still, they created a desire for more, by producing new ways of seeing and new attitudes toward the seen.

But the key to our interest in preserving and packaging imagery came from that first branch from the camera obscura: photography. For in addition to the various forms of projection and visual spectacle for public viewing, the camera obscura led also to the fixing and preservation of images. Projection for public display, combined with visual capture for private use, would eventually rejoin in the form of the projected motion picture—but first we must trace the origins of photography.

## Origins of the Photograph

Photographs and the technologies that make them have attracted amateur enthusiasts for well over 150 years and have served science and commerce for even longer. The photographic camera has roots in the camera obscura and the understandable desire to fix and copy that mysterious image on the wall of the darkened room or its miniature equivalents. It also has a complex history that gradually transformed photography by two distinct ways: first, by deskilling picture taking, and second, by radically reducing the time required to capture an image. The former led to the amateur camera and the informal snapshot (think Kodak), the latter to the motion picture centrally distributed in distinct packages to consuming public audiences. Let's explore.

While the camera lucida made it possible to trace landscapes and public scenes, it remained a crude and limited art. In fact, the French J. N. Niépce's (1765–1833) ineptitude in drawing and tracing led him to seek an alternative way to reproduce an image from nature. His experiments in "sun writing" drew on the long-known fact that photosensitive chemicals turned dark when exposed to light, making it possible to create a negative mirror image, which could then be reversed for an exact positive copy. Beginning in 1816, Niépce placed a light-sensitive coating of silver chloride (later bitumen) on a metal sheet in a miniature camera obscura to create a vague image of a scene when the pinhole was opened. The problem was that it took eight hours of exposure and Niépce had no means of "fixing" the image, of stopping the chemical reaction to light when the image was sufficiently exposed. In 1829, Niépce agreed to collaborate with Louis Daguerre, the inventor of the diorama. Following Niépce's death in 1833, Daguerre developed an entirely different method of fixing images. In 1835, he treated Niépce's silver-coated copper plate with iodine just before placing it in a miniature camera obscura (exposing the plate for four to ten minutes). The result was a latent (not yet visible) image. Then, by exposing the plate to mercury fumes, he revealed the positive image. In 1837, he solved the final problem when he found that a solution of salt in hot water would fix the image, halting the reaction of the photo-sensitive material to light. He called the resulting metallic image a daguerreotype, and in 1839 gave the technology to the public in exchange for a French government pension of F6,000 per year.[16]

Photography took off with astonishing speed. One day after Daguerre's invention was made public in August 1839, a manual describing the technique was published, with editions available in all major European languages by the end of the year. Within a month after its announcement there was a daguerreotype studio in New York City, and inside a year the new technology had forced many portrait painters

to become photographers. To increase light gathering and shorten required exposure, Daguerre's followers improved the lens (typically
making it larger—up to twelve inches); others even hand-colored the
finished daguerreotypes for demanding customers. The daguerreotype quickly won public acceptance because Daguerre's first photos
were unique and irreproducible, in conformity with the expectations of
paintings or drawings. The daguerreotype also offered a personal image
at relatively low cost and without much waiting, compared to sitting for
a portrait. However, because the exposed copper plate had to be developed quickly, daguerreotyping had to be performed and developed in
professional studios, many of which sprouted along the main streets of
towns large and small across the United States and Europe.[17]

While the daguerreotype captured and commercialized the desired
personal image, it was really only the first of four significant photographic technologies, each of which had its own distinctive impact on
producers and consumers. Between 1839 and 1855, the daguerreotype
of silvered copper competed with alternatives such as the positive tin-
type and ambryotype. Next, from 1855 to 1875, a wet plate process
produced negatives on glass or iron sheets covered with collodion (a
derivative of guncotton) into which photosensitive compounds had
been impregnated; these negative images were reversible and reproducible on paper, making possible multiple copies of positive images
for mass distribution. Between 1875 and 1895, a faster-acting dry plate
process involving a gelatin coating on glass replaced the messy and
slower-acting collodion. Finally, after 1884, paper-based roll film, soon
replaced by celluloid, allowed a series of shots one after another that
could then be factory processed altogether. This final development set
the stage for two dramatic innovations: the amateur snapshot camera
and the motion picture.[18]

Technologies often have curious and circuitous origins. Photography is an interesting case in point. Collodion came from the 1846 discovery of guncotton, a new explosive derived from raw cotton treated
with nitric and sulfuric acids. When guncotton was dissolved in di-
ethyl ether and alcohol, it became collodion, a wet gelatinous substance
that served admirably as a photographic substrate. In 1851, Frederick

Archer, an English sculptor and amateur photographer, found that collodion spread on glass plates served as an excellent carrier of halogen
salts, a photosensitive material, which when treated with silver nitrate
just before exposure produced a latent (not yet visible) negative image.
When subsequently treated with ferrous sulfate, a visible negative appeared consisting of dark areas of opaque silver wherever light had
struck the glass—the more light the darker. After chemically fixing the
image, the photographer would then prepare a sheet of paper coated
with a silver halide salt like silver chloride—which, when put under the
negative and exposed to light, yielded a positive image. Thereafter the
print was fixed by washing it in sodium thiosulfate.

This was a cumbersome and messy process compared with the daguerreotype; it also produced a grainier image and required a professional photographer from beginning to end, precluding any kind of
spontaneous or amateur photography, because the picture had to be
taken while the collodion was newly prepared and still wet. There were,
however, some advantages to the new process. The exposure time was
reduced, but the key advantage was that unlike the daguerreotype, the
collodion produced a negative, making its photographs reproducible.
All of this contributed to a craze for cheap personal photographs, especially during the Civil War. Photographers, using collodion as well as
other processes, set up makeshift studios in army camps, producing
thousands of images of dashing lads in uniform to be sent back home.
These images created a much-cherished visual link between anxious
families back at home and the young men torn from them in war, many
of whom would never return.[19]

This new possibility of photography on an unprecedented commercial scale led to two very different innovations of great social import:
the carte de visite and the stereograph. From about 1854 on, photographers began to mass produce photographs of celebrities from negatives. Commonly called cartes de viste, these 4×2½-inch photographs
put the likeness of Lincoln, Queen Victoria, and eventually many other
famous men and women into the homes of both the rich and common. Staged images in bust or full-length poses had flooded the market
within a few years, offering consumers a radically new feeling of con-

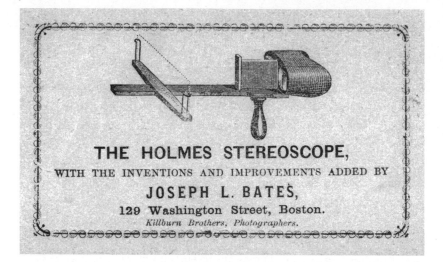

**Figure 6.5** Oliver Holmes's Stereoscope, a simple way to display three-dimensional images (stereographs), from the early 1860s. Library of Congress (LC-DIG-ppmsca-08784).

tact with their heroes. By effectively fanning out the image of the one to the many, the carte de visite helped to create our modern cult of the celebrity.[20]

The second breakthrough in the collodion era was the stereograph, invented in 1840 by Charles Wheatstone of England. The stereograph recreated the visual equivalent of three-dimensional sight—the impression of depth—close to how humans naturally experienced the world. Wheatstone's images were originally of drawings, but by 1849 David Brewster had developed a stereoscopic camera. Brewster photographed a single scene with two cameras about the same distance apart as the human eyes, thus producing slightly different pictures. The two resulting photographs, when viewed in a stereoscope simultaneously but with each eye seeing only one image, reproduced the natural stereoscopic vision of the human brain. At first merely a scientific curiosity, stereographic photographs were commercialized when collodion made possible mass-produced images in the 1850s. Early stereoscopes were cumbersome and expensive, but the American doctor Oliver Wendell Holmes—father of the Supreme Court justice by the same name—

invented a cheap and convenient hand-held alternative in 1861. Customers could even purchase pornography on stereograph cards along with respectable images of tourist sites and children's stories. Like the carte de visite, stereographs became collectables. Only after 1900 did they lose out to the more "exciting" images of the movies, but the stereograph, like the magic lantern, survived as a toy for several generations of twentieth-century children (older readers may remember the Viewmaster).[21]

Like early sound recordings, the first photographs captured what people wanted to preserve — the rare, the remarkable, the unrepeatable, the evanescent. Professionals took pictures of hieroglyphic inscriptions and Egyptian monuments or cityscapes, along with castles and cathedrals of old European towns. Seldom did very early photographers notice common people or scenes. Instead, many from the 1850s sought to reproduce the visual effects of romantic landscape painters with dreamy, even blurry images of gnarled trees and forlorn ruins. At first, Daguerre assumed that only buildings and nature could be photographed because of the long exposure time required (three to fifteen minutes). Improved daguerreotypes made photographic portraiture more attractive, but even so, subjects were still most often presented to the camera in a staged and formal pose, given how long one had to maintain a rigid position. Early photographers insisted that photography strictly reproduced nature, leaving painters to claim that the photograph was merely objective and unimaginative.[22]

Enthusiasts, however, went in a different direction. To them photography offered more than objective truth. It created infinite reproducibility. Holmes, in 1859 in the *Atlantic Monthly*, effused that form was "henceforth divorced from matter"; there was "only one Coliseum or Pantheon; but how many millions of potential negatives have they shed — representing billions of pictures?" The unique creative object or its artistic representation was unimportant. Holmes dreamed that "the time will come when a man who wishes to see any object, natural or artificial, will go to the Imperial, National, or City Stereographic Library and call for its skin or form." He went so far as to suggest that "perhaps in ten years or so the question will be seriously discussed . . . whether it will be any use to travel now that you can send out your artist to bring home Egypt in his carpetbag to amuse the drawing room with."[23] After all, the eyes pressed against the stereoscope saw exactly what eyes saw in the heat, noise, and dust of Egypt without the travail of travel. For Holmes, Egypt was reduced to an accurate and easily accessible image.

Photography taught mid-nineteenth-century moderns to embrace a positive attitude toward rapid technological change. The photo, produced by advances in chemistry and optics, was magical and liberating. It did not threaten a way of life (except perhaps for a few painters). Some even claimed that the photograph enhanced individuality, allowing consumers to portray themselves personally and quasi-permanently to others. According to Holmes, the carte de visite and stereograph made "all mankind acquaintances"—at least in the superficiality of the captured image.[24]

This leads us to the third and fourth stages of nineteenth-century photography. Replacing collodion with gelatin-based photography, first on glass, then on flexible celluloid rolls, allowed the camera to be taken into new places and put into the hands of new classes of people. Richard Maddox, an English physician, in 1871 discovered that silver bromide salts mixed with a gelatin emulsion on glass plates (and later on paper), formed an excellent "dry" substitute for wet collodion. By 1873, a gelatin emulsion was available commercially to professional photographers. The advantages were plain. Gelatin-coated glass plates could be prepared long before exposure, eliminating the need to wet-coat the plate immediately before taking a picture. This saved time and effort, because dry, factory-prepared photographic glass could be carried almost anywhere, allowing pictures to be taken on the go—making possible amateur photography. This dry process also reduced the time of exposure, enabling more informal and frequent shots. The factory could even develop your pictures, removing another set of onerous tasks in photography, while fostering a centralized corporate photographic industry in tandem with the amateur photographer. One of the key beneficiaries of this transformation was George Eastman, who in 1883 began mass-producing prepared photosensitive glass along with facilities to develop and print photos for both amateurs and professionals.[25]

The fourth and final phase in the rise of the modern camera was the development of celluloid roll film, the key innovation that led to the snapshot camera, mass amateur photography, and the entirety of celluloid cinema. Indeed, the next major innovation in photography would not come until the rise of digital photography, in the final decades of the twentieth century. Here's how it happened.

Working with William Hall Walker, Eastman in 1884 replaced glass with roll paper as a backing for the photosensitive gelatin emulsion. This still-cumbersome procedure was improved in 1887 with a simplified process, facilitated by the development of rolled celluloid to replace paper and a new camera, the Kodak. Nitrocellulose, formed by nitrating cellulose, was transformed into celluloid when cooked under pressure in camphor—with the result sliced thin for rollable film. First sold in 1888 to the public for $25, the Kodak included a roll holder with sprockets for feeding edge-perforated celluloid film across the shutter to expose a series of images. The first Kodaks could take up to a hundred pictures; when fully exposed, the photographer sent the camera to the Eastman Kodak factory for developing, printing, and reloading.[26]

Eastman's genius was to develop an extraordinary system that centralized all aspects of photography apart from the picture taking itself, making this new art available for mass consumption by untrained amateurs. From the 1890s, through massive advertising in the new glossy magazines, Eastman taught Americans that they could take their own pictures using easy-to-load film and factory processing. Marshaling a vast retailing and servicing network, Eastman anticipated other complex delivery systems, including those later developed to market the phonograph. The *Kodak Primer* (1888), written for novice photographers, makes this achievement very clear. All the amateur had to do was to "point a small box straight on and press a button." No need for "soiling the fingers" or investing in "exceptional facilities." The oft-heard advertising slogan summed it up: "you push the button, we do the rest." But Kodak still offered advice about ideal settings and subjects for picture taking: wedding trips, summers in the Catskills or at the seashore, trips abroad, visits to California, even winter vacations in the South. Eastman wanted the Kodak to become an essential part of an emerging American leisure culture.[27]

Magazine ads stressed that the "Kodak system" meant ease of use and, as if to reinforce the point, insisted that cameras no longer required male expertise but could in fact be enjoyed by women and children. In the 1890s, this especially meant the "Kodak Girl," the young woman off on a holiday. And after 1900, even children were urged to take up the hobby.[28]

Kodak ads also touted continuous "improvements"—such as faster shutters and film for indoor snapshots and action photos. Eastman Kodak even sold album books to store and display photos, just as the Victor Talking Machine Company marketed album storage for their records. And again like the phonograph industry, Kodak offered a full line of cameras. Though the company's first model from 1888 cost a hefty $25, sales rose sharply with the introduction of the $5 Pocket Kodak in 1895.[29]

Paralleling the marketing strategy of Victor and Edison phonographs, Kodak constantly upgraded its cameras for the upscale consumer while simultaneously expanding its customer base by down-marketing simpler models in hopes of reaching the aspiring middle and working classes. This now-familiar marketing scheme reflected a distinctly American income structure: a smoothly rising pyramid of wages—as opposed to the sharply demarcated plateaus more typical in Europe. But this appeal to a "full line" of consumers also encouraged an American Dream—the expectation that everyone could participate in a "democratic world of consumption" while also marking individual achievement—by "rising up" the scale of consumption as one grew older and climbed the economic ladder. The same marketing technique soon applied to cars.[30]

Eastman's advertising was about more than selling a camera and convenience. It was also about teaching Americans to "see" differently. As historian Nancy West notes, Eastman Kodak sold the "charm" of photography and tried very hard to make Kodak into a verb: "kodaking" meaning looking for the delightful image to photograph. This resulted in more cameras and film sold, but it also created a snapshot vision that has shaped our modern way of seeing.[31]

By placing the camera in the hands of the amateur rather than the professional, Eastman created a whole new way of thinking about and taking photographs. First, he fostered a culture of plenty: think how shocking it must have been, in 1888, to encounter a camera capable of taking a hundred photos at a time (even though rolls of smaller capacity soon prevailed)! Even for people in the middle class, it was rare to have more than a handful of photos. Cheap and convenient photography led to the "stockpiling of memories," with snaps either neatly labeled and dated in an album or thrown into an old shoe box and half-forgotten until after grandma's death. Accumulating photos was part of a recent drive to possess and collect objects ranging from dolls and antique furniture to stamps and baseball cards.[32]

Kodak also tapped into new sentiments about family and time, breaking with older notions of what a photograph should be. Since the 1840s, professional photographers had specialized in formal and iconic images of special places and persons—like Niagara Falls or Lincoln—but also posed images of family that were designed to evoke idealized notions of sex and age roles. As a special and relatively expensive family investment, photos were supposed to be timeless heirlooms. Family photos typically featured authoritative images of fathers, with youth and especially children posed in subservient roles or dressed up, sometimes awkwardly, for rite-of-passage events, such as graduations. Another photographic subject was the sentimental (or to us, morbid) "mortuary photo" of the newly deceased toddler, who might be posed sitting upright in a chair, frozen in idealized memory. In the 1840s and 1850s, mortuary photos were more common even than pictures of the living child.[33]

Through its advertising and instruction books, Kodak invited its customers to break from the professional photographer's clichés of picture taking by rejecting the Victorian ideals they embodied. Nancy West identifies two phases of this transformation. First, by encouraging the young female photographer, Kodak called for a new theme of picture taking, capturing the playful in adults, especially women. Young women were instructed to take their Kodaks with them on their carefree outings and vacations, where they would capture not just iconic scenes but those special personal moments of fun among friends. "A vacation without a Kodak is a vacation wasted," claimed a 1903 ad. Kodak's promotion of the spontaneous snapshot was to reveal a personal truth that the formalized and stylized professional could never achieve. No longer were people to be reduced to an idealized or stereotyped status, age, or gender role. And what made the individual who they "really were" was to be revealed best in those informal moments, especially in times of leisure and relaxation. This really quite curious notion reveals new trends, including the increasing embrace of self-expression and the moral necessity, indeed the obligation, of leisure.[34] The snapshot aesthetic meshed with the growing informality of courtship—marked by the rise of unchaparoned "dating" culture.[35] Kodak promised that pictures taken during the fun and romantic phase of life would allow married couples later to "keep all these precious, fleeting moments alive forever . . . in sparkling pictures." A Kodak ad of 1930, showing a couple on a summer boating adventure, evokes this image: "hours pass, days end, streams run their course . . . but the pictures Kodak gives you are yours forever." The tagline reads: "Romance never fades—in snapshots."[36] Kodak repeatedly evoked this notion of capturing the fleeting—romance and vacations, but also just the special ephemeral moment, whether that be a sports event or even the rare sight of a fish or bird.[37]

The second phase of the transformation was a desire to capture the image of the growing child—perhaps by those "liberated" Gibson Girls who had taken cameras on romantic vacations in the 1890s but who had, by the 1900s, become mothers. These women abandoned the morbid professional photos of dead children and embraced the snapshot of the happy and energetic baby and toddler. This newfound celebration of the young arises as a new cult of the "cute child" replaced older images of young children as angelic or as pitiful urchins. The snapshot image of the delighted and delightful child emerges about this time, in part because parents expected their children to live, given the decline infant mortality. In these images of the cute, adults also projected their own desires for pleasure and fulfillment onto the "wondrous" child.[38]

Kodak reinforced this snapshot aesthetic by encouraging children to become photographers as well. In 1898, Kodak began marketing a cheap, small camera for kids: the Brownie. Like Edison's Gem phonograph, the child's model cost little ($1 for a camera that allowed only six exposures), and was advertised to parents as simple enough for any

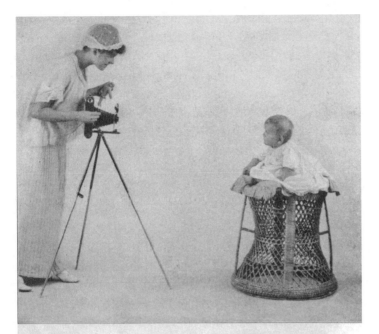

*Keep a*

# KODAK BABY BOOK

THE first journey downstairs for exhibition to that secondary consideration—father. The toddling nursery days! That all important epoch when *the* baby first trudges off to school. In all these great events are limitless opportunities for the Kodak.

And with the school days come pictures *by*, as well as pictures *of* the children. Pictures they take of each other, free from constraint or conscious posing. Spontaneous pictures that reflect simplicity and weave into the Kodak Book the touch of naturalness.

### EASTMAN KODAK COMPANY,

*Ask your dealer or write us for free illustrated booklet. "At Home with the Kodak."*　　　ROCHESTER, N. Y., *The Kodak City.*

**Figure 6.6** A typical Kodak ad instructing the mother to capture those special events of childhood with photography. Children are also called upon to take "spontaneous pictures" that provided a "touch of naturalness" for a Kodak Baby Book. *Saint Nicholas Magazine*, September 1915, 20.

school child to operate. But Kodak also appealed directly to kids. The company organized "Brownie Camera Clubs" for the young and associated its kids' camera with a popular children's illustrated story series about a team of nocturnal childlike fairies, or "Brownies," not unlike today's Smurfs or Teenage Mutant Ninja Turtles. These familiar figures appeared regularly in early Brownie ads in magazines like *Youth's Companion* and *Saint Nicholas* as well as (then) family periodicals like *Cosmopolitan*.[39]

Kodak's celebration of wondrous innocence revealed more than a new cultural fixation on the child or the replacement of formal (professional) by informal (amateur) photography. It was also about freezing the all too fleeting in an image whose emotional power could be evoked every time the photograph album was opened. Sweet sixteen "comes but once in her lifetime," a Kodak ad from 1911 reminds us. Again and again, Kodak insisted that the taking of amateur pictures was about the "pride of firsts." This heavy-handed ad from 1928 summarizes an appeal that long preceded it:

> What a thrill I got as I watched my two boys play their first games of baseball. . . . Overnight my boys became young men. . . . I began to realize then how . . . those snap-shots I had taken were now more precious than rubies. Incidents that otherwise would have faded from my memory I am now able to recall clearly and joyfully. . . . Some day you will want to remember your children as they used to be. When that day comes are you going to be one of the unfortunate few who have no pictures to remind them of life's most precious moments?

When Kodak began producing motion picture cameras for home use in the late 1920s, a similar theme appeared: "They are boys and girls so short a time. . . . There is so much you want to keep: Junior racing down the drive, astride his Kiddie-car . . . Jane . . . strutting about in her first high-heeled shoes. . . . Deny yourself no longer the thrill of living over again your children's happy hours." By the end of the 1920s, Kodak ads even feature "memories" of Grandma: "every little gesture and expression . . . so much herself . . . so REAL." Capturing grandma, though, was almost an afterthought: the focus was to be on the wondrous but fleeting innocence of childhood, the hopeful progressive future.[40]

The snapshot camera transformed the sensual through its ability to fix the ephemera, offering new ways of thinking about time and memory. Philosopher Roland Barthes characterizes the photograph as the "absolute particular," but even more it fixes or stops life even as it claims to preserve it.[41] Whereas Victorians had used the "eternal" photographic image as a bulwark against the terror of time, Kodak reduced the anxieties of change in a new way. It captured in the snapshot the "timely" image, thus making it timeless—a joyous instant rather than a frozen icon. The new ideal was not to commune with the afterlife in a symbolic world of universality as suggested by the Victorian photo but to capture some particular moment of pleasure and love. This took form in two distinctly modern values: the sensual romance of the "date" and the wondrous innocence of childhood. Kodaking, like earlier photography and portrait painting, was to be an aid in avoiding painful memories. But now, only happy scenes were to be photographed. And photography became part of an elaborate effort to ignore, even transcend and deny, death. For many in an increasingly secular age, the photograph was a kind of solace.[42]

By capturing the immediate and informal, snapshots addressed another quintessentially modern anxiety: accelerating change. Kodak reminded its customers that change was inevitable, but then promised to help freeze time at its most wondrous moments. The captured ephemeral in the photo served to domesticate the future, shaping it in highly personal ways. The snapshot aided memory (as ads incessantly pointed out), but the whole Kodaking process was about the future: the whole point of taking and curating photos was to make it possible to "recollect" these pictures later. This anticipation of the future channeled what was seen and shot, as the photographer in the search for the spontaneous and personal ended up basically shooting according to Kodak's instructions. Spontaneity turned into a new kind of ritual dictated by Kodak's vision of appropriate nostalgia. It's a strange notion when you think about it, this act of looking for future memories to capture in the present—a kind of robbery of the present for the future.

The camera and the snapshot (again, like other packaged plea-sures) created a new kind of sensibility, capturing fleeting moments of brothers and sisters at play, couples in love picnicking, or dimples on the chins of babies, and making these pictures into "reality." The snap-shot reflects a shift away from a culture that valued "eternity" in deeply symbolic and unchanging images. The new goal was to capture delight rather than death, even if the Kodak camera also aided in the reduction of experience and memory to a kind of private vision. Kodak did a lot more than develop and print pictures.[43]

## Going to the Movies

Capturing the singular image was only one way to package sight. The idea of trapping and reproducing life in motion for many would be the ultimate prize. Technology and desire combined, and with hindsight we can recognize that most of the essentials for movies were in place by 1850: technologies for the projection of transparencies (as in the magic lantern); methods for exploiting persistence of vision in the sequen-tial display of figures (as in the zoetrope); and the ability to chemically fix and later reproduce images from a miniature camera obscura. Still missing was a chemical process fast enough to allow for the rapid taking of photos in sequence and the mechanism for taking and displaying such a sequence. A solution came in the form of two related break-throughs of the 1880s: the gelatin-based photosensitive "film" that re-duced time of exposure from minutes to only a fraction of a second; and the flexible celluloid roll film of indefinite length, capable of capturing a close sequence of images. This series of transparent images could then be viewed—like the zoetrope—but in a far longer series, creating the magical illusion of motion.

The dream of a moving picture has roots in scientific curiosity, the desire to capture events that happen so fast, that they are invisible to the unaided eye. Eadweard Muybridge, an English-born photographer who settled in the American West in the 1860s, was hired by California tycoon Leland Stanford, founder of Stanford University, to prove that a galloping horse could have all its legs off ground for at least the blink of an eye. In 1878, using a series of cameras to photograph a racing horse in rapid sequence, Muybridge proved just that. Two years later, he developed a circular glass plate with a sequence of action figures drawn around the edge, which could then be spun to project an impression of motion. Another pioneer was the French scientist Étienne-Jules Marey, whose interest in bird flight led him as early as 1873 to devise a photographic "rifle" to record a sequence of images capturing the motion of a bird's flapping wings. His first efforts were for naught, but in 1890, after shifting to faster gelatin plates and then to celluloid, he succeeded. Marey had little interest in commercial applications, but his experiments with a roll film projector were improved in Paris by Louis and Auguste Lumière in 1895, resulting in one of the world's first motion picture projectors to be used for popular amusement.[44]

Meanwhile, back in America, Thomas Edison, witnessing the advances of Muybridge (who visited Edison's West Orange lab in 1888), decided to make "an instrument that does for the Eye what the phonograph does for the Ear, which is the recording and reproduction of things in motion." Using profits from his phonograph, Edison gathered a team, including William Dickson, to design a motion picture camera and viewer. Naturally, he modeled this effort after other "playback" devices of his time, the rotating disc (like that invented by Muybridge) and especially his own cylinder phonograph.[45]

Always trying to enhance his earlier inventive packages, Edison experimented with lining up some forty-two thousand tiny images on a cylinder in spiral form (like his cylinder phonograph) with an eye piece newly attached for viewing. But the impracticality of this device, plus his encounter with Marey's experiments in strip film at the Paris exhibition of 1889, led Edison to take another tack. Late in 1890 he switched to roll film for motion pictures, following a much earlier model, that of the stock ticker strip and printing telegraph that he worked on as a youth. By May 1891, Dickson had a sprocket-feed mechanism adapted to advancing perforated celluloid film, similar to the Kodak camera's film roll marketed three years earlier.[46] The resulting Kinetograph adapted the continuous motion of a crank or electric motor into the intermittent progression of the film roll, matching the opening and closing of the shutter, thus exposing a rapid succession of individual photograph frames. The relatively slow film available at that time required very bright light. To make practical movies, Edison's staff built the so-called Black Maria Studio in New Jersey in 1892. This fifteen-by-fifty-foot structure, resembling a police wagon with the same nickname, was covered in black tar paper and painted black inside and could be turned on a pivot to follow the sun, in order to maximize light. After developing, the completed motion picture could then be played back in a Kinetoscope—to be viewed by a single person. This peep show–type device was a box through which passed a forty- to fifty-foot strip of developed film between a magnifying eyepiece and an electric light.[47]

Following closely on the model of his successful coin-operated phonographs, Edison sold Kinetoscopes for use in arcades, hotels, and similar public places, and manufactured brief novelty films for private "peep hole" viewing. In 1894, Kinetoscope parlors erupted across the urban landscape just as photographic studios did a half century earlier. The craze began in April of that year at a Broadway site, where the owner paid $2,500 for ten machines. Upon entering the parlor, viewers would be greeted by an elegant bust of Edison, which the inventor himself deemed undignified and ordered removed. Soon thereafter, bars, drug stores, and even department stores sported Kinetoscopes, as did seaside resorts, where crowds gathered to see the new and novel machines. By January of 1895, video parlors had even reached outposts like Butte, Montana, and Waterloo, Iowa. *Edison Phonographic News* summed up the initial reaction, comparing the experience of the Kinetoscope to "Aladdin and his Wonderful Lamp": "Now-a-days any man who possesses a nickel can produce more wonderful results by dropping it in a phonograph or Kinetoscope cabinet than Aladdin could in summoning a Genii by rubbing his lamp." Seemingly anywhere and at any time, the consumer could call forth that mysterious power.[48] In fact, that nickel bought only a twenty second "hit," perhaps long enough to set an intelligible scene but still short enough to whet the visual appetitive so the consumer would want to see the films in adjoining machines. Parlors often offered six movies for a quarter.[49]

Much like penny candy (sometimes dispensed from similar coin-op machines), these quickie movies provided an instantaneous thrill. But the kick wore off quickly. Kinetoscopes were available only from 1894 to 1900. They cost a hefty $250 (with films adding another $5 to $8 each) and were profitable only for eighteen months or so before people just stopped coming. The polished wooden box of the Kinetoscope was the classic packaged pleasure of the 1890s, but it was no more than a transitional novelty.[50]

The idea of projecting the motion picture onto a larger-than-life screen was probably inevitable. After all, for centuries audiences had enjoyed the magic lantern and, for years, the panorama and diorama. In 1895, the motion picture projector was invented independently by the Lumière brothers and the Latham brothers (the latter in association with Will Dickson, who left Edison to form the Mutoscope and Biograph Company—soon to be Edison's main American rival in the movie business). That same year Thomas Armat and C. Francis Jenkins invented yet another version and sold it to Edison. Each of these had different devices to move the film through the projector.[51]

These first big-screen movies were typically placed as short novelties in vaudeville houses and circuses. Other motion pictures appeared in back rooms of penny arcades and storefronts, festooned with bright posters to draw in customers. Not all venues were plebeian. Films that appealed to the middle class, especially of travel and news events, played in exhibition halls and even churches, along with upscale magic lantern programs. They were typically less than ten minutes long and were often less sensational than the colorful mechanical illusions offered at worlds' fairs and amusement parks like Coney Island (see chapter 7). Films were part of a broader culture of spectacle, and the film and railroad simulations of Hales Tours anticipated the now familiar video "rides" common at Disney and Universal amusement parks.[52]

These early films of the 1890s were certainly a curious assortment, appealing especially to immigrant and working-class audiences. At first, the motion pictures of Edison, the Lumières, and the Mutoscope and Biograph Company were little more than visual novelties, showing women feeding doves, a dancer with gown fluttering, or even just close-ups of a kiss, men leaving the factory, or familiar New York City street scenes. Soon there were simple stories, but mostly on prosaic topics like a barroom scene where a cop expels a couple of toughs in a fight. Not surprisingly, mildly burlesque scenes like Fatima's belly dancing (famous from the Chicago fair of 1893) were popular, as were boxing matches.

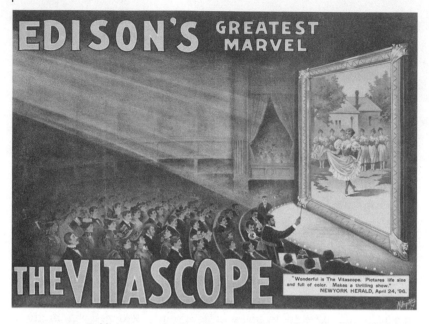

**Figure 6.7** Publicity poster for the projected motion picture via Edison's Vitascope. Note the sophisticated audience and fantasy image of the motion picture as a painting, apparently to appeal to an upscale crowd. Metropolitan Print Company, 1896. Library of Congress (LC–DIG–ppmsca–05943).

Early movies drew also on traditional festival and sideshow fare: cock fighting, "ratting" (a dog trained to kill rodents in a ring), fighting cats and somersaulting dogs, the Cotta Dwarf from the Barnum and Bailey circus, and bodybuilder Eugen Sandow. One film even featured a fake political debate between presidential contenders Grover Cleveland and Benjamin Harrison that ended with comical blows between them. Filmmakers also drew on a common desire for the distant and adventurous. Edison and others offered a wide variety of exotic scenes in short films like *Bucking Bronco*, *Annie Oakley*, *Opium Den*, and *Paddle Dance (of Somoans)*. All offered the visual kernel only: the story of the lives of Queen Elizabeth and Mary, Queen of Scots, long portrayed on the stage and in books, was reduced to the trick photography of Mary's beheading—an act completed in only twenty-two seconds.[53]

How do we make sense of these strange films? Partly they are about "sights" that a viewer might have heard about but couldn't see because of cost, distance, or the passage of time. Many appealed to a working-class sensibility, being mildly sexual, mocking authority, and with action featured over stories. A common thread was the novelty of viewing that which normally could not be seen, or you would not want to see in person: a supersized version of someone sneezing right in your face, or a tooth being pulled in a dentist's chair, or a gruesome execution.[54]

Gradually, however, the movie broke free from the confines of the variety show and sideshow spectacle. By 1903, regional film exchanges had emerged to rapidly distribute ten-minute, single-reel films to mostly fly-by-night storefront exhibitors. Over the next few years, these small venues were transformed into nickelodeons. Still relatively small with room for seating between fifty and six hundred and screens usually only nine by twelve feet, nickelodeons became common near streetcar stops, subway exits, and main streets. Identifiable by a recessed entry and box office, the nickelodeons would often have flashing electric lighting and barkers or piano players to lure in the passing crowd. There may have been as many as ten thousand nickelodeons in the United States by 1910, many of which were owned in chains. The program usually ran continuously throughout the afternoon and evening, offering a quick fix of entertainment in an informal setting for crowds on the go.[55]

In 1890, the French sociologist Jean-Gabriel Tarde observed a change that would well describe the impact of the cinema a few years later: the spectacle experience was being transformed from the seasonal, local, and unchanging event (as in the festival or fair) to a nearly daily, perpetually changing entertainment shared by viewers across the nation and even the globe.[56] Media historian Tom Gunning points out that film was originally not so much an "extended dramatic performance with unified plot and action"—but rather more like a hit of packaged pleasure.[57]

What perhaps best distinguished early films (between 1894 and 1907) was their devotion to the visual thrill, what Gunning and others have called the "cinema of attractions."[58] Early cinema was about showing the spectacular, not telling a story. It offered visual surprises (magical tricks or physical feats), displays of the exotic or grotesque, natives, scantily clad females or freaks, and speeding trains or explosions. Trick movies used techniques like film splicing to make objects or people appear or disappear, sometimes creating realistic illusions (as in films of executions or the explosion of the *Robert E. Lee* steamer). Crowds learned from the tricks of magic lantern exhibitors to expect to see ghosts and apparitions, but with film these illusions were made more graphic by the double exposure technique used in *Uncle Tom's Cabin* (1903) to create the fantasy of angels carrying off the departed. Fairy tale magic was popularized in Georges Méliès's famous *A Trip to the Moon* (1902), *Kingdom of Fairies* (1903), and *Wonderful Wizard of Oz* (1910). Cartoon drawings moved fantasy through scenes that could not be realistically shown (as in Méliès's man landing on the moon). Winsor McCay used cartooning to create whimsy in films like *Gertie the Dinosaur* (1914) or for dramatic sequences in the propaganda film about the sinking of the *Lusitania* (1915). The first movies thus offered new visual perspectives, not unlike hot air balloons, microscopes, and spyglasses when they first appeared. Cinema forced consumers to see differently.[59]

By the beginning of the twentieth century, however, this visual novelty was not enough. Filmmakers began to supplant tricks and side-show hits by "actualizing" events appearing in newspapers and magazines, prefiguring the newsreel. William Dickson and his partners at the Mutoscope and Biograph Company made about five hundred "actuality" films between 1896 and 1902. Like Georges Méliès and Edwin Porter, Biograph filmed travel themes, especially the famous phantom runaway train movies, but also reenactments of the Spanish-American and Boer wars. Movies of celebrities were produced, featuring the likes of Kaiser Wilhelm, Cecil Rhodes, and Pope Leo XIII. Chase films became popular with "action pictures," such as Edwin Porter's 1903 *The Great Train Robbery* (twelve minutes of gun play, fast trains, a heroic girl, and final retribution). And humorous stories like Biograph's nine-minute *The Suburbanite* (1904) told of the travails of a middle-class family's move to the suburbs, ending with their decision to return to the city. These films remained snippets of events, animated photographs really, and closer to the new genre of Sunday comic strips than theatrical plays.[60]

These early movies were about the packaging of the visual, examples of what media scholar Henry Jenkins calls the "wow complex," shared by many commoditized sensualities of the time.[61] The movie enlarged the possibilities of sight, creating perspectives formerly accessible only from air balloons, fast trains, or expensive holidays. Commercial films eventually picked up on the theatrical tradition of storytelling, but the early focus on thrills of action and the visually unexpected never really died out. Indeed, we see a return of "action movies" in the 1970s, with the gut-wrenching staged "beats" of intense, violence-filled, sequences in *Dirty Harry*, *The Wild Bunch*, *Shaft*, and *Star Wars*, and wildly popular action-adventure films starring the likes of Arnold Schwarzenegger, Jean-Claude Van Damme, Chuck Norris, and Bruce Willis.

## Movies after 1907

Historians of the cinema identify the years after 1907 with the rise of the narrative film. This trend coincided with the decline of cheaply made shorts shot around New York City and played at nickelodeons and the rise of feature films produced in Hollywood for increasingly grand movie theaters.

Let us take a quick look at what changed and why in the decade after 1907. This period started with Edison's domination of the American movie industry and ended with his eclipse. In 1907, Edison finally won in the courts his claim that Edison patents on the movie camera and projector were being violated by all but the Mutoscope and Biograph Company, forcing those errant companies (including Vitagraph, Selig, Lubin, Essanay, Kalem, and Pathé) to pay license fees for using equipment that had infringed on his patents. Meanwhile, in 1908, Edison with his licensees joined with Biograph to create the Motion Picture Patent Company (MPPC). By pooling patents, these companies attempted to create a legal monopoly to control all production, distribution, and exhibition of movies in the United States. The idea was to force everyone to use only licensed MPPC equipment—and to show only films made on that equipment. For a time, George Eastman joined with Edison's group, agreeing to sell film stock only to MPPC members (though the availability of European-made film stock made difficult this attempt at monopoly). The MPPC also replaced sale with rent of prints, thus controlling copying, quality and duration. Based on a model established by Edison in his record business—the regular and timely output of entertainment—the MPPC required that films be standardized: marketed by the foot, usually changed three times a week, and sold in bundled fifteen-minute programs with no advertising or promotion.[62]

However, as was true in many consumer goods industries around this time, the movie business became an oligopoly, not a monopoly. After much litigation, a U.S. Circuit Court of Appeals declared Edison's MPPC an illegal restraint of trade in 1916. Within two years, Edison, Biograph and virtually all other MPPC members were marginalized or out of business. Replacing them was a group of independent moviemakers and distributors who had fought the monopoly and won—and ended up creating what we now know as the major film studios of Hollywood. Distributor Carl Laemmle formed Universal Pictures in 1912, for example, and theater chain mogul William Fox created Fox Film in 1915. Adolph Zukor, founder of Famous Players in 1912, joined with Feature Players to form Paramount Pictures in 1914. And the three Warner brothers—Jack, Sam, and Harry—who started with nickelodeons in the mining towns of western Pennsylvania, moved to Hollywood in 1917 and founded Warner Brothers formally in 1923. Other significant movie mills (including MGM, Columbia, and United Artists) emerged slightly later. And although it would be simplistic to say that the MPPC failed to innovate—a claim often made in regard to Edison's record business—the rise of these independents certainly coincided with dramatic shifts in movie plots, movie audiences, and the whole idea of movie stardom.[63]

Those changes had roots in the successes of the nickelodeon era. The insatiable demand for new motion pictures, which the MPPC could not meet, guaranteed markets for independent producers. The movies, after all, had become a daily habit for millions, and exhibitors had to learn how to keep feeding demand as film production rose from 678 titles in 1908 to over 5,000 by 1913. Movies also became longer, requiring more than a spectacle, gag, or a chase scene to attract an audience. And with this, films began to tell stories. The nascent film industry drew on the narrative traditions of popular entertainment: the print or performed melodrama, the dime novel, and the "Blood and Thunder" sensationalism of the cheap theater of the 1890s. Freelance writers of formulaic stories from these high-output industries joined the movie makers (much as the recording industry used the talents of Tin Pan Alley song writers). By 1910, western and Civil War stories from pulp fiction were being turned into screen plays, creating a uniquely American genre of patriotic and moralistic films;[64] and after 1913 serials or chapter movies were shown successively over many weeks to draw regular customers.[65]

As part of this change, movies developed increasingly complex plots and characters. Unexplained events were eliminated, and plot conflicts tended to get resolved by the end of the story. By 1903, "intertitles" were added to explain the action and provide necessary dialog in this still-silent era. New techniques of close ups and angle shots, combined with artful film editing, also moved the story along far more quickly and efficiently than had previously been possible in an era of live theater.

To explain more fully why moviemakers embraced the narrative of the play, we need to look at changes in who was consuming film and who was censoring it. The success of the nickelodeon era in creating a mass, regular audience holds the key. This crowd of moviegoers (consisting mostly of young people) made middle-class moralists anxious about what was being seen. This led to repeated efforts to censor films (as in Chicago in 1907 and New York in 1908). Rather than fight the censors, exhibitors formed their own New York (soon to be National) Board of Censorship in March 1909 to weed out films that a churchy middle-class viewer might find lewd or glorifying crime. In fact, exhibitors were generally happy to submit to censorship in order to gain more upscale customers (who could of course be charged higher admission fees). Moralistic melodramas flourished in this environment, leading to the films of D. W. Griffith and Cecil DeMille, who adopted theatrical narratives (appealing to the affluent) without sacrificing the thrills and excitement of the cinema of attractions (pulling in working people). Romantic stories were combined with battle and chase scenes, for example. Innovations such as these helped to create a cross-class movie culture quite distinct from that found in European motion pictures from this era.[66]

Along with the narrative film and middle-class tilt came the feature film and the "movie star." This changed the packaging of movies in several ways. A lot of early film producers could not imagine an audience willing to sit through anything longer than a single fifteen-minute reel. Breaking from this system, some of the independent distributors began to offer full-length theatrical films from Europe. In 1912, Adolph Zukor exhibited the French four reeler *Queen Elizabeth* to movie houses that could attract a more affluent audience. *Queen Elizabeth* was notable as a theatrical play–length "feature" and also for its featured performer, the aging (but still famous) Sarah Bernhardt. By contrast, Edison and other patent company movie makers had made no effort to identify any of their actors. The Italian-made epic on the Roman Empire, *Quo Vadis*, took forty-five minutes to view in a 1913 American version, cut down from its original three-hour span. Its success prompted the production of longer and grander motion pictures, especially Griffith's Civil War epic *Birth of a Nation* (1915) that lasted an astonishing three hours and fifteen minutes.

Along with the feature-length film came the featured performer, the star. This innovation is often associated with Carl Laemmle's decision in 1910 to identify Florence Lawrence as the lead actor. Laemmle also paraded her before the public, at one point even faking her death to juice up her notoriety. Other actors, like the curly haired Mary Pickford, quickly gained a following (and sold movies) with personal appearances and magazine interviews. The "star system" of course had precedents—for example, in vaudeville—as a way of distinguishing one show from another. Stars did very much for films what brand names did for canned or bottled goods. But in the movies, "stars" also emerged from audience expectations. Regular patrons of early movies swooned over and idolized the personalities of actors; many eagerly followed their lives off stage in the new genre of fan magazines (the first was *Photoplay*, founded in 1911). Attending their favorite actor's movies gave moviegoers a sense of being intimate friends with their idols, and the more they knew about their roles, personalities, and lives, the more they enjoyed the movie.[67]

A final feature of the new video package was the movie palace, built to attract an affluent audience that often was put off by the plebeian air of

**Figure 6.8** A classic from the era of the movie palace, the Granada Theater on the north side of Chicago, circa 1933. Note the ornate organ alcove. Library of Congress (HABS ILL,16–CHIG,109–17).

nickelodeons. In the 1910s, chains of large, elegant, and often "themed" movie theaters emerged in the downtown shopping hubs of cities or in residential areas near streetcar stops. These movie palaces retained the visual clues of vaudeville and theater, with vestibules, recessed box offices, protruding marquees, and colorful posters and tracer light displays. Many could seat a thousand or more patrons—some in New York could seat 5,300. The high-end houses included organ accompaniment (rather than a piano) and employed ushers dressed in braided uniforms with epaulets and featured interior décor in mock Spanish Renaissance or Oriental themes. Movie receipts rose from $301 million in 1921 to $720 million by 1929, four times more than all sports and live theater sales combined.[68]

## A Packaged Way of Seeing

By the 1920s, the motion picture (along with the snapshot camera) was wildly popular in the United States. Films, modular and prepackaged, were reproducible and easily distributed across thousands of movie houses—with the marginal costs of an additional projected show close to zero. It is hardly surprising, then, that movies dramatically lowered the price of popular entertainment, by comparison with live theatrical plays. This made the regular movie habit possible for millions, even as its producers also made their millions. Vaudeville and other theaters featuring live performances gradually disappeared in all but the largest venues—like New York City—and even attending the American pastime of baseball remained more or less a special event, not part of the everyday or weekly routine.

By 1920, popular photography and movies were relatively cheap and accessible, but they also intensified and extended the universe of what ordinary people could see, liberating them from the natural limitations of premodern optical experience. Both were born in science but later became the objects and instruments of popular and commercialized pleasure. Recall that the camera obscura led to both personalized amateur photography and the commercialized moving image projected on a screen for a paying audience. And as the camera became faster, more portable, and easier to use, people learned to look for less formal, more personalized gestures, and to accumulate these images in unique family photo albums. The snapshot became a way of manufacturing personal nostalgia. Picture taking was now in the hands of the consumer, but the manufacturer continued to "package" the photographic vision, influencing what was seen and captured in the snapshot.

A distinctly modern way of seeing is even more true for the movies. As Walter Benjamin famously noted, the motion picture liberated people from the natural constraints of sight, training them to see and react to changes in their lives and to imagine themselves in new roles and circumstances. The movie camera "can use certain processes, such as enlargement or slow motion, to record images which escape natural optics altogether," and motion pictures had "the unique ability to use natural means to give incomparably convincing expression to the fairy-like, the marvelous, the supernatural." This power of the camera lets the viewer "discover the optical unconscious, just as we discover the instinctual unconscious through psychoanalysis."[69] Or so he thought.

Packaged sight preserves and extends our vision as it focuses and looks into places we could never see on our own, and is potentially liberating. Yet it also distorts what and how we see. The motion picture had special ways of wowing the audience (as did the panorama and magic lantern before it). It not only went where the eye couldn't normally go—up close as no one could comfortably experience in real life, top-down views as if the audience was a bird in flight, and into a fantasy world (making it visually real). Still, the most striking feature of the packaged visual event is that it could be packed—compressing time and thus intensifying the visual effect. As psychologist Hugo Münster-berg famously wrote in 1916, the movies sped up time. Here is how he described an adventure film: "Within the span of fifteen minutes the buccaneers scuttled a merchantman, made its crew walk the plank, captured a fair-haired maiden, bound her with what appeared to be two-inch Manila rope, and cast her into the hold," a series of events that would have taken two hours in a "Third Avenue home of melo-drama" and no doubt much longer in real life. Münsterberg concluded that movies were capable of "stirring up of desires together with their constant fulfillment."[70]

Maybe Benjamin was right that the motion picture's almost tactile, "percussive effect on the spectator" was how modern people learned to adapt to a confusing and rapidly changing array of stimulations. Movies train us to cope with this assault on the senses. But the *visual* wow, extending and intensifying what we see, furthers a long process of privileging sight over other senses as well the sceptic over intellectual reflection and the cultivation of other emotions. While stories and characters did not disappear as theater advocates feared when viewing early movies, the stress on visual intensity often reduced the role of the word and of character development, especially in many films since the 1970s. And hasn't the wow dulled our sensibility toward the greens, browns, and grays of nature and the pace of a gathering storm or a sunset?[71]

For intellectuals like Benjamin, the mass-produced and packaged image in the photograph of the Pyramids or a painting by Rembrandt has lost the "aura" of the original[72] even if photos may be the only way that most people can ever "visit" these places or see these things. For Holmes in his stereographs, this was inevitable, even desirable. He was content with the "form" freed from the "material" reality of an object. Yet certainly something was lost in the reproduction even as so much was gained.

And still we purchase and consume the "form" because it can be packaged and can be packed with sensuality often missing in the original. The result is the spectacle, as in the cinema of attractions, and it has become ubiquitous. Perhaps Guy Debord goes too far when he says, "The spectator feels at home nowhere, for the spectacle is everywhere." But the visual spectacles have displaced the everyday, and if we haven't become exactly addicted to it, we sure spend a lot of time behind the lens and in front of the screen. This our ancestors learned to do in the early days of the Kodak and the movies.[73]

**7**

# Packaging Fantasy

*The Amusement Park as Mechanized Circus,*
*Electric Theater, and Commercialized Spectacle*

The pleasure packs we have considered so far served to preserve, portion out, compress, and eventually label and commercialize some biological desire or sensation, meeting some need rooted in human biological or cultural history. Packaging often transformed those needs by overcoming some natural scarcity, at times creating excess and disease or some other, more subtle, psychological or sociocultural effect.

Yet another type of packaged pleasure with equally long roots can be found in this same *fin de siècle* era: the multisensual commercial pleasure site. So here is our final package, those self-contained places that offered bright lights and garish colors, along with unusual smells and tastes and loud and cacophonous sounds and recreational sensations of bodily motion, all compressed in time and space: the amusement park.

Unlike the other pleasure packs chronicled in this book, the amusement park was a site, a confined space. It captured a multiplicity of sensations, becoming a kind of sensorial emporium. At times, this sensorium encouraged sociability rather than isolation—unlike some of the other packagings we have encountered. The amusement park was nonetheless a "packaged pleasure" in the way we are using that term.

And while its most famous form took shape first at Coney Island on the southern tip of Long Island, not five miles from Manhattan, this compressed collection of spectacles, novelty foods, exotic entertainments, and thrill rides was recreated, essentially mass produced, in thousands of similar parks all across the world. And, as a distinctly mass commercial and technological package, the amusement park redefined two age-old human creations: the local and time-bound festival, and the elite pleasure garden.

From an anthropological point of view, the festival can be seen as a natural outgrowth of the collective sharing of food after a successful hunt, a sociable rite with ecological advantages. It typically took the form of periodic gorging, a sociopsychological comfort in societies where scarcity was the norm and surpluses could not be easily stored for later use. Festival rituals, especially communal dancing, must also have given vent to protoreligious emotions, further helping to bind groups together. Ecstatic dancing with rhythmic music probably facilitated human bonding long before chiefs, armies, and states coerced "community."[1] Festivals became more elaborate in agricultural and urban cultures, often doubling as trade fairs, intensifying, ritualizing, and giving meaning to bits of sensuality. Festivals added new sensations—like the physical exhilaration of sport or mock combat, combined perhaps with recreational vertigo from swinging and falling, thrills that eventually would be mechanized in the modern amusement park ride. Regulated by a religious and/or agricultural calendar, ancient and medieval festivals periodically transformed sites of worship or work into places of pleasure and wonder.

Pleasure gardens—ancestors to what we more commonly think of as "parks"—were another root of the modern amusement park. As distinct sites of sublimity, discovered in nature but improved by human hands, pleasure gardens (meaning not for food or grazing) may have originated with some kind of nature worship. With the containerization of surplus and creation of private wealth, however, pleasure gardens became prized and often exclusive refuges of the rich and powerful: sites of status, fashion, and the personal whims of their owners, with the growth of enclosures that privatized many forms of previously public space. In Europe in the seventeenth century, aristocratic or royal pleasure gardens started to become commercialized and opened to a paying but progressively more plebeian crowd.

Of course the general phenomenon of compressed, commercialized entertainment can be found in other cultural institutions: the circus, the theater, the opera or concert, the puppet show, the sporting competition, and numerous others. In the nineteenth century, yet another (and more immediate) ancestor of the modern amusement park was the "world's fair," an invention of Victorian scientific and commercial optimism first organized in London in 1851. Like the pleasure garden, the world's fair promised genteel uplift. Over time, however, such exhibitions were blended to accommodate the plebeian appeal of the festival in a more commercialized pleasure-garden setting. From these diverse roots emerged a distinctly new and more mechanized form of entertainment: the amusement park.

## Fairs and Festivals

Holidays and the festivals that accompany them have an interesting environmental and economic history. In temperate zones of human habitation, late-autumn and early-winter festivals have tended to coincide with the in-gathering of crops, fulfilling also the need to slaughter livestock with the end of the pasture-growing season. Holidays have tended to coincide with seasonal surpluses—and lulls in agricultural labor or the irregular demand for artisan work. Festivals were often held at the site of grain or meat surplus or some sacred or holy space where part of that surplus might be sacrificed to the gods to assure next season's harvest. The English holiday season, for example, extended from All Saints' Day (the day after Halloween) through Christmas and New Year's, ending with Plow Monday (the first Monday after the twelfth day of Christmas), when farmers were supposed to return to work the fields. Another holiday phase marked the winter downtime and the blossoming of spring: Shrove Tuesday (Mardi Gras) was followed by the midwinter Lenten lull of austerity until Easter and Whitsuntide (Pentecost), times of spring festivities. Long even before the rise of monotheisms, these festivals of springtime renewal were celebrated with colored eggs (in tenth-century BCE China, for example).[2]

Feast days of course were food days, times for satisfying pleasures of the palate. But they also very often marked extremes of the sun's movement north and south in the sky, bringing and taking away life-giving warmth. Feast days were thus often days celebrated with religious and calendrical rituals. Greco-Roman and Semitic peoples divided their year at January and June (near the solstices) and September and March (the equinoxes). The Celts, however, bisected it at the first of November and beginning of May, coinciding with the end and beginning of summer pasturing of herds. The first day of May, of course, was important for other Europeans, who made it into a day marking the power of fertility. May Day typically commemorated spring with a May Queen and pole around which villagers danced and sang with brightly colored streamers and decorations. The early medieval church co-opted pagan winter solstice feast days, turning them into Christmas. The Celtic Samhain (the New Year beginning at nightfall on October 31) was likewise transformed into All Saints' Eve, or Halloween. Midsummer (June 23–24) was rechristened the Feast of St. John, in honor of John the Baptist, though pagan rites survived, including midnight jumping over bonfires and bathing in rivers. These were sometimes acts of purification and attempts to assuage anxieties about the fate of the upcoming harvest. Most European holidays had at least formal ties to feasts honoring saints or to the liturgical calendar.[3]

Festive times often coincided with annual trade fairs — gatherings facilitating the exchange of goods. London's St. Bartholomew's Fair was a ten-day event beginning August 24; that fair lasted from 1131 until 1855, finally succumbing to Victorian standards of order and decency. Fairs such as St. Bartholomew's offered sundry merchandise as well as a colorful array of entertainments, facilitating consumption of hard-to-find goods for societies not yet blessed with retail shops. Long before the circus sideshow, rural fairs featured traveling wax figures of the famous and notorious, paintings of historic events, and "freaks" such as dwarfs, pig-faced ladies, and people from exotic places. Hiring fairs brought young men and women together, providing opportunities for heterosexual encounters in evening dances.[4]

Festivals offered more than trade or coupling, however. They also involved a pleasurable release from drudgery, routine, and social con-

**Figure 7.1** William Hogarth's image of the frivolity and chaos of London's Southwark Fair, 1732. The fair was an annual August event that was shut down by the authorities because of riots in 1762. Library of Congress (LC–USZ62–120284).

straint—nicely summed up in the word "Saturnalia," the ancient Roman festival of mid-December, honoring the god of plenty, Saturn. Preindustrial Europeans also very often binged during festivals, partaking of food, drink, sex, or even fighting. In southern Europe, Carnival was the most common saturnalian festival, though similar elements can be found in many other holidays—English Christmas or the New Year's Mummers Parade in Philadelphia, for example. Carnival was characterized by its Latin root, *carne*—meaning meat, with implications also of the flesh or sex. People indulged in food and drink, but Carnival was also a time for the beginnings of premarital pregnancies.[5] Normal constraints of civil and hierarchical life were very often lifted at such events, as revelers threw flour, fruit, eggs, and even stones at each other and sex roles were transgressed (men worn women's clothes and vice versa). Bishops and dukes might even be mocked in plays and in songs.[6]

It is probably not overly psychologizing to say that saturnalian festivals served as a release for people who daily endured the rigors of routine and scarcity and the humiliation of having to submit to authority. But if festivals turned the "world upside down," they also took place in confines of space and time set by tradition. And because they released psychological and social tension or just exhausted participants with excess, authorities very often tolerated these festivals, at least until modern times. Participants could be counted on to return to work or to accept the rigors of a more or less formalized period of restraint—as during Lent, originally a time of anticipated food shortage.[7]

An oft-told story is the effort of religious purists and then secular reformers to suppress Saturnalia. From the seventeenth century, European authorities regulated or even withdrew financial support from traditional festivals, fearing their excess and challenge to authority both sacred and secular. By the nineteenth century, reformers attempted to create more polite alternatives to the rowdy festival. Victorian innovations included not only what we now think of as public parks, incorporating leisure activities like rule- and referee-based games such as soccer and cricket but also child-focused holidays such as Christmas and Halloween. But Saturnalia survived in festivals like the modern Mardi Gras of New Orleans, spring break bashes in Florida, outdoor concerts of various sorts (especially from the 1960s on), New Year's parties, and parties surrounding sporting events of various sorts—World Cup or Super Bowl parties, for example.

Festival culture was also perpetuated in the industrial and urban setting of the early-twentieth-century amusement park—albeit with significant changes. In place of traditional aggressive games and the mocking of authority, modern saturnalians sought the thrill of mechanical rides and the playful celebration and self-mockery of modern machines. Instead of the cyclic return to a set of carnival traditions, the amusement park offered an action-packed experience often based on the latest news or fad, compressing a former time's intensity of sights, sounds, smells, movement, and crowds into a new type of sensual commodity.

## Pleasure Parks and Amusement Parks

Fairs and festivals are one source leading up to the modern amusement park, but another source was the pleasure garden, a forerunner also of our modern urban park. This too has ancient roots, dating at least as far back as the exclusive and often landscaped hunting preserves of the kings of thirteenth-century BCE Assyria, grounds stocked with game or exotic beasts for the royal sport of trophy killing. Even more spectacular must have been the famous Hanging Gardens of Babylon (604 BCE), a set of irrigated terraces and walls covered with exuberantly flowering plants and trees, all carefully selected for the private enjoyment of local elites. The Hanging Gardens and other Babylonian parks formed such a contrast against the surrounding barren plains that some scholars believe them to have inspired the Garden of Eden story. Centuries later, the ancient Greeks turned forested groves into shrines or even imagined homes for their gods. Wealthy Athenians constructed urban parks as refuges from the throngs and for recreation. The Roman general Pompey built public gardens in 55 BCE that befitted the magnificence of the capital's center. Most parks of this sort, however, remained exclusive. Roman palaces would often incorporate private imperial gardens featuring decorative grottoes, olive groves, irrigated terraces, and artificial ponds, copied later by Renaissance Italian princes and French kings. Following Roman precedent, medieval European aristocrats created hunting preserves that combined natural beauty with the thrill of the chase.[8]

Landscaped hunting parks evoked thoughts of paradise, plenty, and power. Such was supposed to be the impact of the orderly gardens that surrounded the mosques of medieval Moorish Spain. Louis XIV combined these themes at Versailles, in the 1680s and '90s, when he turned a royal hunting lodge into a lavish palace with a grand "backyard." Louis's garden, designed as a half star nearly ten kilometers across, was bisected with radiating straight avenues and canals interspersed with terraces, pruned trees and hedges, parterres, groves, and sculptures—with an extensive hunting preserve beyond. This was a striking display of royal power over man and nature, but it was also a setting of sensually rich fêtes for royal guests—with parades, feasts, theatrical performances, fireworks, and dancing. The gardens became a backdrop for colorful displays of novelty and grandeur, anticipating the spectacles Disney would present more democratically in his theme parks.[9]

Quite different was the English garden of the seventeenth and eighteenth centuries, favored by the local aristocracy. Drawing on Chinese fantasy landscapes, the English garden often included grottoes and artificial streams adorned with romantic bridges, pagodas, and crags (or even contrived ruins). Abandoning the geometric regularity and formalism of the French garden, English landscaping was characterized by a quasinatural, pastoral, and often idiosyncratic (or romantic) presentation, suggesting to many English writers the values of "Anglo-Saxon freedom" in contrast to French absolutist displays of power through forced geometric landscaping. English gardens, however, were in fact made possible when powerful aristocrats confiscated common lands, depriving the poor of access to pasture for their animals and to wood lots for fuel, tools, and construction. And English gardens were anything but haphazard. Much effort went into creating an impression of the "natural" and romantic. And the English, too, transformed the garden into a site for spectacle. In the eighteenth century, Henry Hoare created a pictorial circuit garden at Stourhead, south of Bristol, fashioning scenes drawn from the ancient Greek tale of Aeneas's journey, including miniature temples to Apollo and Hercules, a collection of ponds ("Diana's Basin"), and a grotto with statues of nymphs. Like modern "theme parks," Stourhead was supposed to provide an escape from everyday life, treating the visitor to a timeless "away place" with a kind of story that, like an emplaced myth, could now be inhabited rather than just imagined or recited.[10]

Landscape historians often find in pleasure gardens a kind of openness and narrative subtlety missing from the modern commercial theme park.[11] But this may underestimate the continuities between the two sensorial sites. Pleasure gardens eventually came to feature a combination of spectacle and nostalgia: "The terrifying and giganticized nature of the sublime is domesticated into the orderly and cultivated nature of the picturesque," as Susan Stewart describes the contained vistas of the pleasure garden.[12]

With the growth of the urban bourgeoisie in seventeenth-century

England, a new type of pleasure garden appeared—accessible to city-dwellers and open to the paying public. Along with rural healing water spas and later the seaside resort, this commercial pleasure garden became a refuge from urban congestion and from work for the aspiring classes. Expanses of green space, manicured gardens, fountains, and alcoves were set in semi-exclusive environs accessible to those able to pay a modest admissions charge.[13] Commercial pleasure gardens also offered a more civilized alternative to the rowdy and unkempt festival while retaining many festive elements; they also allowed for festival-like experiences all or much of the year, providing opportunities for a "visit on demand" for those with time on their hands and money. These pleasure gardens appealed to a more affluent crowd, some of whom may have only recently withdrawn their patronage from saturnalian festivals. Despite their exclusivity, these gardens and their cousins, the spa and seaside resort, offered experiences that ultimately shaped the commercial amusement and later theme park.

Just as the gallery painting and the salon concert framed, refined, and ultimately privatized (and commercialized) the traditional sights and sounds of the communal celebration, the new type of pleasure garden privatized and, for a time at least, commercialized the traditional festival. The best illustrations are the gardens of London—especially Ranelagh (opened in 1742), Marylebone (from the mid-1600s), and Vauxhall (laid out in 1661 but popular only from about 1727). Small gardens were also built around mineral springs or on the lands surrounding coffee- or teahouses—of which there were as many as sixty-four in and around London at the height of their popularity. Most of London's pleasure gardens had graveled paths, lawns, clipped hedges, grottoes, fountains, and statues. Some had arbors for tea drinking. The better-funded had large buildings for concerts and a bun or cake house. Swings and roundabouts (carousels) also appeared in such parks. Historian Roy Porter observes that "admission fees were great [social] levelers,"[14] at least between the aristocracy and the rising middle class.

Vauxhall, a twelve-acre site south of the Thames, was probably the most famous, winning the patronage of literary and artistic notables like Oliver Goldsmith, William Hogarth, Joshua Reynolds, Jonathan Swift, and Henry Fielding. In the evening, a small orchestra and organ

would be positioned at the center of a grove with surrounding pavilions, where groups of six to eight people would sit to dine and listen
to a concert of songs and short popular pieces. But the entertainment
was often less genteel—including fireworks, jugglers, tight-rope walkers, and "salt-box music," with masked performers playing rolling pins,
broomsticks, and other mock instruments. Over time, the pretensions
of early pleasure gardens gave way to the tastes of the carnival crowd.
Vauxhall's owners had to accommodate the herd in order to win new
customers. Novelty acts were introduced, such as the 1827 live reenactment of the Battle of Waterloo, complete with a cavalry charge. Urban
decay eventually caught up with Vauxhall, however, and the site was
shuttered in 1859. The affluent were driven away by an increasingly
seedy surrounding neighborhood and by park crowds of ever-lower
"quality." These were common problems in the pleasure garden business. Urbanization of adjoining real estate and the drive to attract ever-
larger paying crowds made it difficult to maintain a high "social tone"
over time. Today, even Disneyland in Anaheim shows signs of succumbing to a similar fate.[15]

New Yorkers imitated London with a long string of pleasure gardens—including five called Vauxhall Gardens (the first appeared in
1767). These were often attached to inns, with their lives cut short by
high-density land use as the city grew. Crowds were attracted with an
eclectic array of statues, *trompe l'oeil* paintings, and an ever changing menu of music and special events (including balloon ascensions,
horse shows, tightrope walks, and necromancers, often entertaining by
colored lanterns at night). As early as 1805, a Vauxhall Garden in New
York reenacted a scene well known to readers of contemporary newspapers: the U.S. Navy's defeat of the Barbary pirates at Tripoli who had
captured an American frigate off the coast of Libya, the first great overseas engagement by the U.S. military. The USS *Intrepid* was depicted
in miniature in a daring raid of the pirate base while the enemy fired on
the invading Americans from a castle and battery on shore. The drama
concluded as the city of Tripoli went up in flames while the American
invaders escaped in triumph. All this involved elaborate mechanical
models, painted scenes, fireworks, and an audience seeking a patriotic
thrill. Later in the nineteenth century, however, these spectacles would

**Figure 7.2** A perspective of Vauxhall Gardens, 1751. William Wroth, *London Pleasure Gardens* (London, 1896), 300.

suffer the same fate as those in London. They could no longer attract an upscale crowd, and New York pleasure gardens became associated with the rough culture of their Bowery environs. Manhattan elites would later abandon these spectacles for the "natural landscapes" featured in the planning of Central Park, opened in 1859.[16]

Pleasure gardens for a time shared certain affinities with the spas that had emerged around "healing waters." The practice of bathing in thermal mineral springs dates back to prehistoric times and was common in the ancient world; in Medieval Europe, however, such bathing was generally confined to holy sites (associated with saints) and reserved for healing the sick. Such was the role played by the English town of Bath, which only became a center for fashionable leisure during Elizabethan times.[17] Other sites closer to London, like Tunbridge or Sadler Wells, became important as "drinking spas," where the wealthy would socialize in "pumping rooms" while drinking local mineral water (for their health). These spas typically offered dancing, music, and quiet spaces for reading, along with access to an attached garden.[18] Like the regular pleasure gardens, these spas gradually shifted from aristocratic display and genteel reserve (albeit with liberal doses of hypocrisy) to plebeian-pleasing spectacle. Sadler Wells, for example, featured panoramas and, from 1804, its theater specialized in aquatic shows, becoming a music hall in the 1870s.[19]

Seaside resorts like Scarborough began as competitors to inland spa springs, offering seawater to drink and bathe in and promising cures for gout and worms. Brighton on the southern coast of England, however, developed into something less rigorous and more fun. After 1750, it became a thriving social center for the idle rich, especially after winning the patronage of the Prince of Wales. Particularly noteworthy was the prince's construction of the iconic but eccentric Royal Pavilion in 1786. Brighton and other British seaside towns became places for daytime promenading and evening concerts and dances, as piers formerly used for docking boats were decked out with pavilions for entertaining crowds flowing from the beach promenade.[20]

Many of the newly wealthy in the American colonies, especially after independence, emulated their homeland in this aesthetic. Coastal towns like Newport, Rhode Island, became summer retreats, attracting visitors from as far away as the Carolinas with its Gulf Stream waters and cooling summer breezes. Inland, White Sulphur Springs and other Virginia health resorts followed the decorum of Bath. Northern resorts like Saratoga Springs attracted health-seeking crowds to its mineral waters but quickly adapted to pleasure seekers with promenading, drinking, gambling, and, from 1863, horse racing. Later in the nineteenth century, the east end of Coney Island with its Oriental and Manhattan hotels and private railways from New York City, provided an upscale social setting that was isolated from the less polished crowds to the west on Long Island's Brighton Beach, but still flourished on horse racing and firework extravaganzas until the 1910s.[21]

In the United States, many sublime sites of nature were also transformed into commercial spectacles. Niagara Falls, for example, was turned from a venue for cultured contemplation of God's natural world to a commercialized spectacle in the second half of the nineteenth century. The romantic and dramatic vista of the falls brought the crowds, but what kept them there were daredevil tightrope walkers, showmen spinning tales of accidents and rescues, and freak shows exhibiting body oddities like midgets and bearded women. Completely artificial spectacles made these sites commercially viable, intensifying their sensuality beyond what nature alone could ever provide, while displacing or even destroying some of its subtler appeals.[22]

## World's Fairs and Midways

Yet another precursor of the amusement park was the international exhibition, or world's fair. Launched in London in 1851 at the Crystal Palace to display scientific and artistic progress (and national pride), the world's fair also became a venue for novel forms of amusement. And although there were notable European installations of this sort, especially in Paris, the United States took the lead. New American cities showed off their growth and wealth in a long series of exhibitions beginning with Philadelphia in 1876 and followed by Chicago (1893), Atlanta (1895), Nashville (1897), Omaha (1898), Buffalo (1901), St. Louis (1904), Portland (1905), San Francisco (1915), Chicago again (1933), San Diego (1935), and New York (1939–1940). These were consummate displays of the burgeoning consumer culture, offering visitors a kind of "commodity pilgrimage." A catalog of the London fair of 1851 offered an astonishing 1,500 pages of objects for purchase, fueling new faith in the possibilities of abundance after the depressed 1840s.[23] American exhibitions were also notable for catering to popular tastes, and increasingly so over time. A commercialized arena of fun, food, and freaks appeared spontaneously on the edge of the Philadelphia exhibition of 1876, earning it the derisive moniker "Shantyville."

Similar areas appeared in the neighborhood of the 1893 World's Columbian Exposition in Chicago, a stately affair boasting a neoclassical Court of Honor, a lagoon, and sober displays of scientific achievement — with a hefty 50¢ admission charge to keep out the riffraff. Inland and directly west of this imposing compound (dubbed the "White City") was a semiautonomous amusement strip some 600 feet wide and nearly a mile long. This ungated Midway, as it was called, was operated on a pay-as-you-go basis, encouraging a large and socially diverse crowd. Romantic reproductions of a Square of Old Vienna and an

Irish Village with a "Blarney Castle" were included, along with exotic scenes of Algerian and Tunisian villages complete with Bedouin tents, mud-daubed huts inhabited by "warriors" from Dahomey, and a South Sea village featuring "cannibalistic" Samoans. One of the most popular exhibits was "The Streets of Cairo," complete with "Little Egypt," a female belly dancer. There was also a simulated Colorado gold mine and an ostrich farm. Midway exhibits featured whimsical events designed to titillate the crowd with the primitive and childlike. Despite ethnographic and educational pretensions, the show was actually run by Sol Bloom, a 22-year old impresario who also supervised installation of the world's first Ferris wheel—designed by George Washington Gale Ferris—a steam-powered, 264-foot-high ride capable of lifting 1,440 passengers in 36 cars.[24]

Many such innovations started at world's fairs and then were passed on to amusement parks as regular attractions. Buffalo's exhibition in 1901, for example, introduced the Trip to the Moon and reenactments of the Johnstown Flood and Battle of Saratoga, along with scenic rides to hell and heaven that were subsequently reassembled in amusement parks. The St. Louis fair of 1904 offered the Pike, an amusement zone twice the size of Chicago's Midway, which included a scenic railway racing up and down at forty miles per hour and spectacles like the Creation and the Galveston Flood that later appeared at Coney Island and elsewhere. With its postage stamp recreations of foreign cultures and exotic villages, the Pike offered, as media historian Tom Gunning notes, "a world tour compressed in time and space and rendered effortless for the tourist."[25]

The world's fairs taught visitors to look forward to—and to welcome—the new. At the Columbian Exposition, for example, General Electric, still a new company, presented an "electric fountain" with colored lights displaying the magic and power of alternating current. And even if it was fantasy, the St. Louis fair offered the visitor a kind of packaged, simulated tour of travel anywhere. The trip to the "North Pole in Twenty Minutes," with its mock up of an Atlantic steamer, had to be impressive, especially in 1904. Here, in a space of two hundred by fifty feet, a large audience could experience by viewing moving panoramas the illusion of travel from an American city to the arctic, complete with an electric replica of the aurora borealis. Another ride was Under and Over the Sea, simulating a submarine voyage to Europe and an airship return to New York, a spectacle introduced shortly after the Wright brothers' first flight.[26]

World's fairs offered a kind of fantasy or virtual world, a sensual smorgasbord (or microcosm) that no other packaged pleasure could provide. Notable is that the consumer could move about *within* the package, walking or riding through a complex set of attractions that assaulted the senses with the new and striking. As with the early experience of photography, it is hard for us today to imagine how impressive this must have been — and how different from ordinary experience — for Victorian visitors. Physicians warned the elderly or nervous not to attend the St. Louis World's Fair, lest they damage their health from such an overwhelming sensory assault. Implicit also was a kind of training to a new era of commodity culture. Gunning writes that the world's fairs were "showplaces of ever-changing commodity culture," offering "training grounds for a disposable visual culture." Fairs taught audiences the "protocols of modern spectating" and created easily copied thrills that entrepreneurs would soon exploit commercially, especially in amusement parks.[27]

## Packaging Amusement

Amusement parks combined the saturnalian intensity of the traditional festival, the sensual relief of the pleasure garden, and the dense clustering of display and spectacle of the world's fairs — all in a technologically dazzling and commercial form. Like the pleasure garden, the amusement park was a contained physical space within which consumers experienced a multisensory razzle-dazzle. Walking through a virtual maze of delights, the visitor was engulfed by a buffet of sights, sounds, and smells. The experience was very different, though, from the genteel sensibility claimed by the pleasure garden. The boisterous crowds, the unrelenting intensity, and the thrills of bodily motion followed in the traditions of the older festival and the fair, but there were differences. New of course were the electric machines and admission fees, but the amusement park could also be accessed outside the limited temporal scope of the holiday or fair, typically all summer long and sometimes all year round.

Historians often trace the modern amusement park to the Tivoli Gardens in Copenhagen (1843) and Prater Park in Vienna (1866),[28] but it is Coney Island that really ignited an amusement park craze — and that dates only from the 1890s. An oval-shaped protrusion from the southwestern shore of Long Island, Coney Island had long been an escape from nearby Manhattan's business and factory culture, especially from the 1880s. Manhattan Beach on the eastern end of the "island," with its hotel and promenade, promised exclusivity and refined entertainment — like Henry Pain's dramatic firework spectacles, including a reenactment of the defeat of the Spanish Armada. The more downmarket crowd at the center of the peninsula enjoyed Buffalo Bill's Wild West Show in 1883,[29] followed soon thereafter by the spectacular Elephantine Colossus, a 150-foot-high, tin-covered pachyderm. Coney Island was also first to install La Marcus Thompson's switchback railroad, the most important predecessor of the modern American roller coaster.[30]

Coney Island rose to prominence as a site of amusement parks in 1895 when Paul Boyton, famed as an ocean swimmer and inventor of the frogman suit, opened Sea Lion Park with a smattering of rides — the circular Flip Flap coaster, the Shoot the Chutes water slide, and, for couples, a scenic railroad called the Old Mill — along with a dance hall and an arena for performing sea lions.[31] George Tilyou followed in 1897 with Steeplechase Park, named after a ride that playfully imitated a steeplechase race with wooden horses on an undulating track. Tilyou's business was based on a simple idea: "We Americans want either to be thrilled or amused, and we are ready to pay well for either sensation." After a fire in 1907, Tilyou rebuilt his park around a three-acre structure of glass and steel: the Pavilion of Fun. Here he packed an array of frequently changing slides, swings, and other contraptions requiring customers to subject their bodies to twists, turns, and unexpected accelerations as well as to contact with other bodies (in The Barrel of Love, a Trick Staircase, and so forth). Rounding out this pleasure dome were conventional shooting galleries and sideshow attractions.[32]

In 1903, Frederic Thompson and Skip Dundy built a far grander

**Figure 7.3** Luna Park at night, a sight of colored lights and fantastic structures that must have dazzled visitors in 1905 when this photograph was taken. Detroit Publishing, Library of Congress (LC–DIG–ppmsca–10795).

Luna Park on the 36-acre site of the bankrupt Sea Lion Park. The key to Luna's success was the timely exploitation of electric lighting, a new technology simultaneously being used to turn New York's Times Square into a "technological sublime."[33] The new park also featured innovative construction that, along with colored night lighting, transformed Coney Island into a magical site. Here an "electric Baghdad" was constructed largely from "staff," a gypsum-based mixture of chemicals that, when combined with burlap, created a plaster-like substitute for stone. Thompson's concoction allowed him quickly and cheaply to build copies of statuary, towers, domes, and colonnades from ancient European and Asian civilizations. And so by 1911, Luna Park boasted 1,210 red-and-white painted towers, minarets, and domes all outlined in electric lights. Staff turned out to be extremely flammable, however, and there was the added problem of deterioration, which required frequent remodeling to avoid shabbiness. The extremes of class and crudity were expressed throughout the site: at one end of an elongated reflecting pool beset with monumental architecture was a 200-foot-high Electric Tower; at the other end, this stately pool was disturbed by a playful Shoot the Chutes water slide. The rest of the park was a disjointed welter of gaudy, garish, and unexpected structures, decorated by some 200,000 colored electric bulbs.[34]

Luna Park was the brainchild of Frederic Thompson, a spectacle architect with successes at the recent Nashville and Buffalo fairs. Thompson contracted out most of the attractions on short-term leases, producing an ever-changing hodgepodge of circus rings for trained animals, clowns, scenic rides, panoramic mechanical reproductions of disasters and battles, and quaint ethnic "villages." In this sense, Luna Park was more like the clusters of rides and game booths one can still find today at county or state fairs and carnivals. There was little of the thematic unity of Walt Disney's park in California decades later; even so, the site was pioneering in its promise of a constant delivery of sensory novelty. It was really more like a world's fair that went on year after year, always undergoing reconstruction. And all this brought success: in 1904 Luna Park had an astonishing four million visitors, forcing the owners to raise a second deck around the courtyard to accommodate the overflowing crowds.[35]

A third amusement park, Dreamland, was built in 1904 by William H. Reynolds, with the hope of attracting a more upscale crowd on its 15 acres of choice Coney Island waterfront.[36] In contrast to the Oriental look of Luna, Reynolds insisted on a more Western-classical appearance, featuring all white buildings surrounding a stately lagoon. The 375-foot Beacon Tower, modeled after the Giralda minaret of Seville, Spain, stood at Dreamland's western edge, besting Luna Park's Electric Tower in size, illumination, and "class." Even so, this White City look-alike was anything but didactic. Reynolds built not one but two Shoot the Chutes opposite the Tower. Visitors arriving by ferry at night encountered a majestic vista, with a purported 1 million electric lights illuminating the sky. The street entrance was at the site of a ride dubbed the Creation (from the St. Louis Fair of 1904), where visitors strolled under a huge, classical female nude decorated in gold paint. The Creation spectacle held up to its name, treating audiences to an indoor scenic boat ride along a 1,000-foot canal adorned with scenes from the first 7 days of Genesis.[37]

Despite its promised elegance, Dreamland never attracted the crowds from Luna or Steeplechase. Apparently, its attempt at respectability with religious themes and White City architecture failed to appeal to the mostly down-market crowd of central Coney Island. By 1911, the white paint was already peeling off the staff. In the off-season, Dreamland's owners undertook an expensive remodel, abandoning the purity of white for a décor of cream and firehouse red. The renewal was for naught, however, when on May 27, 1911, a fire consumed the entire park inside three hours. It was never rebuilt. The land was leased out for parking and various temporary rides and exhibits, prior to becoming the site of the New York Aquarium in 1957.[38]

These three Coney Island amusement parks—Steeplechase, Luna, and Dreamland—provided a model for countless subsequent theme parks. Especially influential were their innovations in architectural fantasy. Coney's parks (especially Luna and Dreamland) were essentially theatrical sets, designed to evoke "the natural, bubbling animal spirits of the human being," in the words of Frederic Thompson.[39] The towers and rides also added novel perspectives to the architectural panoramic. Riders from the top of Shoot the Chutes, for example, saw a broad swath of park buildings and the Atlantic Ocean before descending on a watery slide into a lake below. All three facilities offered a Shangri-la alternative to the grime of the city. In 1905, Richard Le Gallienne advised visitors that Dreamland should be approached by boat from the sea to avoid seeing its proximity to "squalid neighborhoods and . . . the back of everything" on the railroad ride. A visit to Coney Island at night, especially in the new age of electric lights, offered New Yorkers an alternative to the filth of the tenements and factories and the faceless or even threatening crowds on the street and in the subway; Coney offered an escape into the dazzle of contrived color, light, shape, and motion, a new kind of urban space where crowds could immerse themselves in a new kind of sensorium.[40]

Of course, the commercial point was not just to delight the crowds but to draw them to the ticket booths of the various spectacles and rides. As packages within the package of the park itself, the individual attractions were focused sensoria, offering compressed scenic tours of distant or imaginary places, reenacted disasters, and other dramatic events, even if in the form of a kind of pseudotourism, as historian Daniel Boorstin contends. Panoramas, magic lantern shows, stereographs, and of course newspapers and magazines had stimulated in Americans a yearn for recreational "travel," and amusement parks met this need, even if marred by demeaning stereotypes and artificially compressed decontextualizations of time, space, and place.[41]

Throughout much of Coney Island's golden era (1897–1930), visitors could stroll down the Streets of Delhi—including an Indian palace and processions of horses, costumed soldiers, and elephants. Dreamland's Over the Great Divide (1907) simulated rail travel over the Rockies with an improbable mechanical volcano that would erupt as thrill-seekers journeyed by. Topping this perhaps was Dreamland's Coasting through Switzerland (1904), an indoor scenic railroad outfitted with refrigerated water pipes to keep this confected alpine adventure "as cold and as full of sweet pure air as can be found among the picturesque Swiss mountains."[42]

Coney also gave revelers a chance to travel where no one had ever gone: notably to the moon and to hell and back. Frederic Thompson's Trip to the Moon took the panorama/cyclorama to a new level by combining a full-bodied illusion of travel with a set narrative track. Reminiscent of Georges Méliès's silent film by the same name (from 1902), the attraction combined popular hopes for space travel with children's storybook fantasy. After customers entered a round building to receive a mood-setting orientation from a guide from "the Aerial Navigation Company," who would explain the secrets of "anti-gravitation and aerial flight," the crowd filed through a narrow passage onto the *Luna*, a cigar-shaped moonship. Electric fans created a sensation of forward movement, while painted canvases and projections appeared through portholes to create the illusion of an ascent far above the fairgrounds, as New York City soon becoming only a distant cluster of blinking lights—and earth a mere ball. Stars would then appear through the portholes, along with a luminescent moon. Descending onto the moon through a "sea of sunlit clouds," the ship seemed to land in the crater of an extinct volcano, as passengers were told. The doors then opened and the travelers were greeted by midget "Selenites" — moon natives — who offered the earthlings green cheese. The crowd was then led down a passage bordered by "illuminated foliage of fantastic trees and toadstool growths" to the palace of the Man in the Moon in his throne room, surrounded by dancing moon maidens. While this spectacle seems comical, even ridiculous, to modern readers, note that Thomas Edison and President William McKinley both claimed to have enjoyed the trip. So captivating was the concept that it was copied in Disneyland's Rocket to the Moon a half century later (though without the moon maidens).[43]

The original Trip to the Moon offered a curious combination of appeals, evoking dreams of scientific progress but also a playful, only half serious, romantic fantasy. Dreamland specialized in other kinds of imaginative journeys — including thrilling encounters with the afterlife and underworld. Hell Gate, for example, began as riders in boats descended in a spiral stream where, as one journalist noted, "with your whole soul you combat fear, even transform it into joy" in an underground channel. "Light comes, and with it red devils amid flames, volcanoes spitting fire, gorgeous grottoes all dripping with stalactites. . . . Gradually you retrace the spiral . . . and finally come out upon a little quay, rich in varied grotesqueries."[44] Though tacky perhaps by today's standards, we should not forget that this was a time when hell was still very real and frightening to many. Rides such as this would play with the still-real fears of riders, challenging them to "transform it into joy" by making hell delightful.

The popular quest for dramas of a more heroic/historical nature led to another form of spectacle, the re-creation in miniature of recent battles and disasters. One of the strangest was the 1903 War of the Worlds, seen from a grandstand. Crowds were offered the pleasure of watching a flotilla of German, British, French, and Spanish ships attacking New York harbor (though happily, unsuccessfully). This was followed in 1906 by the Fall of Port Arthur, a reenactment of a dramatic battle in the Russo-Japanese War of 1905, when Japanese torpedoes sank the Russian fleet. Naval themes were also in play in 1908, when Dreamland's Trip to the Moon was replaced by the Battle of the Merri-

**Figure 7.4** A representation of the "Fighting the Flames" spectacle as audiences, many from the tenements, watched a dramatic and realistic portrayal of the burning of a tenement block. *Munsey's Magazine*, May 1907, 558.

mac and Monitor. Equally popular were stagings of the Johnstown and Galveston floods of 1889 and 1900. These shows went far beyond paintings or even projections. A real wall of water burst forth from a replica broken dam, rushing mercilessly down upon a miniature Johnstown, Pennsylvania. A spectacle titled Fall of Pompeii offered replicas of Roman buildings but also new pyrotechnics designed to give an impression of fire belching forth from the interior of the earth.[45]

More dramatic still—and perhaps more frightening—were the life-size reproductions of tenement block fires. Fighting the Flames first appeared in London in 1903, then at the St. Louis Fair in 1904, and thereafter on Coney Island. Four times a day, from grandstand seats, an audience watched the burning of an artificial six-storey tenement building (made of soft flammable wood on an iron frame) from across a street. Involving a reputed cast of two thousand, the show began with a fire breaking out on the ground floor. The hapless dwellers would then panic, climbing from floor to floor to escape the flames. A team of a hundred and twenty firefighters would then arrive on fire trucks to fight the blaze. While many people would escape by jumping into nets, others would flee onto the roof to be saved (by firefighters scaling the walls) just before the roof caved in.[46] Needless to say, there must have been a certain risk to the performers, even in simulation.

This disaster spectacle paralleled film and newspaper accounts of dramatic urban fires and brave firefighters circulating at the beginning of the twentieth century. A version in Chicago's White City amusement park (from 1905) boasted a three hundred by two hundred-foot street scene complete with trolley cars on regulation tracks. The realism extended down to women gossiping behind their tenement windows and businessmen buzzing about in offices before the fire started in a fireworks store. While the buildings of the set were made of fireproof materials, strategically placed flammable materials were placed to "give the proper effect" of a real fire. According to historian Lynn Sally, the idea was a realism that would physically shock the viewer. The show may also have helped assuage the anxieties of people who themselves faced daily danger in their unsafe tenements.[47]

## The Recreational Ride

Beyond all these compressions of sight and sound and the emotions these evoked, the amusement park offered something in addition: the physical sensation of extreme bodily movement, electromechanically produced. The thrilling experience of speed, of rising and falling, sliding and swinging, twisting and turning, evoked vertigo and the rush of a kind of "safe" or virtual fear. But the ride also often produced a joyful release and sense of triumph from transcending anxiety. The cultivated thrill of falling dates long before industrial civilization, as we find from the deep prehistory of bungee jumping, diving, and the like. Young men from the Vanuatu Islands in the South Pacific continue even today the ancient practice of land diving (jumping from high places using vines) as a rite proving manliness, and there is scarcely a child from any part of the world that does not delight in bouncing, swinging, and tumbling. These sensations were surely more common in the mobile, risky, rough-and-tumble life that preceded the settled routines of agriculture and protective cities. In brief bursts of energy and excitement, early humans scampered over rough terrain or even climbed from branch to branch, stretching to reach life-preserving food or life-saving escape. Agilities of this sort must have provided survival advantages, and a proclivity to rehearse them in play may well be inherited. This may be one explanation of the broad attraction of thrill rides. Rides of this sort exploit sensations children seek on playground swings, slides, and carousels; and adults by the end of the nineteenth century were being allowed to enjoy such activities, associating play with a joyful (and newly legitimate) return to childlike amusements.[48]

Of course, though there is a long history of agricultural and urban cultures disdaining thrill-seeking (or risk-taking) for its own sake as wild and foolish, such indulgences were never successfully suppressed (think of the long-standing fascination with gladiatorial or blood sports). Interesting also is the fact that a renewed fascination with the foolhardy "sport" of falling appears about the same time as the visual liberation of the panorama. In 1797, an intrepid parachutist was dropped from a hot air balloon three thousand feet above Monceau Park in Paris, and ballooning itself was part of a growing aesthetic of probing—and eventually commercializing—new and "extreme" physical horizons. Tightrope walking about this time was starting to attract crowds: high wires were walked during the first circuses promoted by Philip Astley in the late eighteenth century, and gravity defiers would henceforth find new venues—and audiences, including many willing to pay for a view. Garrett Soden in his book, appropriately called *Falling*, suggests that "the act of defying gravity was potent in itself because it confirmed the Enlightenment's version of progress; of man as nature's master, not nature's victim."[49]

Early amusement park rides could not offer personalized parachute drops—though a "ride" simulating this thrill did appear at the New York World's Fair in 1939, moving shortly thereafter to Coney Island.

Roller coasters and water flumes generated similar sensations. And the contained thrill of submitting one's body to the vicissitudes of physics went beyond the sensation of falling. At Steeplechase, many rides were little more than adult-sized playground equipment. One such novelty was a gigantic slide shaped like a smoking pipe, assuming that form, perhaps, to hide that it was little more than what children enjoyed on playgrounds. The Trick Staircase, Earthquake Floor, and House Upside Down all appealed to a childlike delight in vertiginous jerking. The Wedding Ring, a wooden circle suspended from a central pole, was likewise basically a children's park swing for seventy adults. At Steeplechase, the Whirlpool and Human Pool Table threw bodies in all directions and often into members of the opposite sex. The thrill also recalled the toddler's pleasure of being tossed into the air by an indulgent parent or uncle.[50]

Designing thrills of this sort accelerated in the early years of the twentieth century, especially as new forms of technology came into being. It was only a few years after the Wright brothers' flight, for example, that Hiram Maxim (of machine-gun making fame) introduced his Airship ride to amusement parks (in 1907) on both sides of the Atlantic. The ride offered only a circular spin in suspended seats but, while "the laws of physics" upheld passengers, as the journalist Rollin Hartt noted, beneath them "the world reels and swells and topples to and fro like mid-ocean billows," giving with each successive moment "a new scale of perspective."[51] Similar effects were achieved in many other kinds of rotating rides, adding a new form of physical intensity to the simple motion of the terrestrial carousel.

The dominant thrill ride, however, was "funneled" on tracks and tunnels in what became the roller coaster. Roller coasters shared with ordinary trains and trolleys the experience of rapid movement through some landscape, but here in the form of a ride it was possible to have all of the boring bits of such trips left out: no dull, unplanned visual terrain or long spells between "hits" of excitement. The roller coaster and its cousin, the scenic railroad, combined the delightful tour with artificially induced vertigo and mock danger.

Roller coasters have multiple origins, depending on what part of the ride one wants to focus on. At least part of that ancestry, though, must trace back to the thrilling descents of those seventeenth-century Russians who, making the best of their long winters, slid down ice-covered hills. This was not at all (in the beginning at least) a child's play, but originally an activity of adults. A simulated hill first appeared in St. Petersburg in the form of a seventy-foot-high structure made from wood, with a steep, fifty-degree incline covered in hard-packed snow down which revelers would slide on sleds. By 1784, a French entrepreneur had built an inclined track using closely spaced rollers (hence the name "roller coaster"—albeit *montagnes russes* in French) on which people could slide in cars all seasons. In 1846, the first big step to the modern thrill ride was the looping coaster in Paris's Frascati Gardens. A forty-three-foot hill provided the momentum to send an occupied cart along a thirteen-foot wide metal arc, relying entirely on centrifugal force to keep cart and rider on the track inside the arc. Elsewhere, coasters had closer ties to railroads. In 1872, an abandoned inclined-plane railroad formerly serving mines near the town of Mauch Chunk, Pennsylvania, was converted into a novelty ride, dubbed the Switzerland of America. Thousands of revelers took the descent nine miles down Summit Hill into the Lehigh River Valley, in open cars bouncing through a series of hairpin turns and bumps, reaching a speed of up to fifty miles per hour by the time it reached Mauch Chunk Depot.[52]

In 1884, the American inventor La Marcus Thompson imitated (and miniaturized) this ride in his gravity-propelled "switchback" roller coaster. The ride was an instant success, thrilling Victorian Americans despite its timid six-mile-per-hour descent. While Thompson's ride originally required a team of men to push the cars up an incline, steam-powered chain lifts appeared in 1885 to raise cars for a rapid, undulating, and twisting descent, and the track became a closed-loop circuit. Thompson added painted scenes along the way to simulate travel to exotic or fantasy places, but it was really the thrilling sensation of vertigo that gave the ride its appeal. In 1895, the Flip Flap coaster at Coney Island turned the rider in a complete vertical circle in a ten-second experience that, for some customers at least, caused neck pains. Soon thereafter improvements in the loop—making it slightly oval-shaped to make the track less stressful on the body—turned this loop-de-loop feature into a thrill ride commonplace (albeit one that still petrifies).[53]

**Figure 7.5** A "looping the loop"-type roller coaster, very daring and dangerous for 1901 in Atlantic City. Detroit Publishing. Library of Congress (LC–D401–13711).

Improved designs made these and other thrill rides ever more central to Coney Island, eclipsing the costly reenactments and other spectacles. Luna Park's Wonder Wheel (Ferris wheel) of 1920 stood 150 feet high and carried 169 passengers. John A. Miller's invention of under-track wheels in 1910 allowed roller coasters to run on higher inclines and sharper turns without having cars jump off the track (a perennial problem that led to a number of catastrophic accidents). In the 1920s, mammoth wooden-frame coasters (on metal rails) were a huge success in amusement parks worldwide, with at least 1,500 operating by 1929 just in the United States. Among the most famous were those at Coney Island, including the giant Thunderbolt, built in 1925, a coaster that was soon superseded by the even higher and more terrifying Cyclone with its 60-degree plunge from a height of 85 feet. To increase the thrill and variety, designers invented an array of new tricks to be played on the minds and bodies of riders—including undulating "camelbacks" and "kangaroo hops," figure-eights, and spiral dips. Roller coasters became the thrill ride of choice because they could become ever more seemingly "scary" and breathtaking with relatively little chance of actual physical injury.[54]

Roller coasters were supposed to thrill and to titillate, transporting riders to the brink of physical danger and sexuality (as bodies brushed and inhibitions disappeared in the excitement). But coasters and other thrill rides were just far enough removed from both to give release and pleasure, rather than anxiety or a break in sexual decorum. Distilling a cluster of physical disorientations, falls, accelerations, and unexpected twists and turns into a few action-packed minutes of motion, the roller coaster isolated and commodified the sensation of physical disorientation. The French philosopher Jean Baudrillard claims that these rides produced a decontextualized euphoria and a "complete exploitation of all possibilities of being thrilled." Certainly roller coasters became a quintessential "hit" of packaged pleasure, compressing as much (mostly) nonsexual physical arousal as possible in as short a time as possible, and always for a price.[55]

The question then arises: why did the thrill ride become such an obsession at the end of the nineteenth century? Thrill rides appear to be part of the same broader process of sensual intensification and commodification we have been exploring in previous chapters. The quest for speed and the romance of instantaneous travel was anticipated in everything from ever-faster trains and streetcars to the visual illusions of the cinema of attractions. And probably for those whose lives were dulled by routine and overwork, thrill rides offered a needed exhilaration.[56] The more sophisticated rides also imitated and parodied the latest mechanical contrivances, simultaneously evoking and assuaging people's fears of them. Luna Park's Tickler spoofed modern urban traffic when up to eight passengers seated in spinning circular cars rolled down a twisting alley, running into posts and other cars along the way. Historian John Kasson remarks on how the undulations of the roller coaster of the 1890s imitated the era's elevated trains, turning the monotony of the daily ride to work into a playful parody.[57] The thrill ride met a particular need in the age of the packaged pleasure.

## Sited Pleasures All Across the Land

Coney Island's parks were imitated in any number of *fin de siècle* set-tings, just as many of Coney's own rides and exhibits began at world's fairs before finding a more permanent home at Luna Park or Dream-land. New parks were typically built at sites large enough to accommo-date the full array of rides and spectacles that consumers had come to expect. By contrast with the downtown shopping and entertainment centers that had begun to emerge slightly earlier, amusement parks were necessarily suburban, located on relatively cheap land. Many were built at seaside resorts or inland picnic parks, but a setting of natural beauty or even comfort was not required. Disney would prove half a century later that the sights, sounds, and other delights within the park itself were more than enough to attract the millions, sometimes making the surrounding landscape and environment irrelevant, or even a dis-traction. Critical was a means of transporting sufficient numbers to the gates, especially during the heavily trafficked summer months. Ferries and other boats were often used but the preferred form of delivery was by rail, especially the electric streetcar,[58] extensive networks of which sprang up in urban centers of both Europe and America in the 1890s. Many amusement parks were in fact originally built by trolley and elec-tric company investors, often at the end of streetcar lines, to increase the draw on weekends when these light rail systems were underutilized. Manufacturers of coasters and other rides and spectacles (the Philadel-phia Toboggan, for example, or makers of fireworks) erected similar clusters of attractions in other parks.[59]

And now we come to Ohio's Cedar Point, today one of the largest and longest-running parks in the United States — and one of the most innovative. At first it was little more than a picnic park on a narrow peninsula jutting into Lake Erie, accessible from Cleveland and Toledo by boat. In 1870, a German cabinet maker, Louis Zistel, built a beer gar-den and bathhouses on the site, but it was only in 1882 that investors succeeded in winning customers with picnic tables, bathhouses, and a dance hall, followed in 1888 with the Grand Pavilion, a theater and con-cert hall, and in 1892 with a switchback railway. In 1899, George Boeckling took control and, following the national trend, made Cedar Point into a full-fledged amusement park. In 1902, the Racer, a figure-eight roller coaster, was built and two years later lagoons were constructed as was customary in the era. In apparent allusion to the Newport swells, a Breakers Hotel was built in 1905. But the next year a new Amusement Circle opened with rides including a carousel and circle swing, as well as a penny arcade and shooting galleries. A Dip the Dips Scenic Railway appeared in 1908 and a much larger Leap the Dips coaster in 1912. By 1920, Cedar Point had three roller coasters, its own native village, a Fighting the Flames show, and even the Eden Musée, a wax museum. Like Disney World much later, Cedar Point featured hotels for the classy and continuous updates of its rides for the masses.[60]

This was hardly atypical. North of Atlanta developers bought a forty-seven-acre site known for its forested scenery, bubbling spring, and lake and called it Ponce De Leon Park (after the seeker of a fountain of youth). A full-scale amusement park was built there in 1903, offering the full buffet of expected attractions: a toboggan slide, a merry-go-round, a Ferris wheel, an Old Mill scenic railroad, a gypsy village, a penny arcade, pony rides for youngsters, boats, and a two-level pier for strolling around the lake. A trolley capable of running five hundred railcars per day delivered thousands of thrill seekers from central Atlanta.[61]

Chicago became another major site of amusement parks, with Chutes Park (1896), Sans Souci (1899) and White City (1905), all on the Windy City's South Side. Luna Park (1908) and Riverview (1904) were on the north near the new elevated trains. White City, modeled after the Pike at the St. Louis World's Fair, included the Bumps (a gigantic slide), a Ferris wheel, a carousel, a roller coaster, a Shoot the Chutes, a scenic railway, a gondola ride called Beautiful Venice, and Hiram Maxim's Flying Airships, as well as Fighting the Flames, Midget City, and the Johnstown Flood, many of which were borrowed from Coney Island. The only attraction with local color was Cummins Indian Congress, a reenactment of the 1812 massacre at Fort Dearborn (which later became Chicago). Like a number of other parks, White City featured a three-hundred-foot-high Electric Tower lit by twenty thousand bulbs and a searchlight. Riverview, though, which by 1907 was at a hundred acres, was Chicago's largest amusement park, attracting nine million visitors by 1911 (compared to White City's two million by 1905).[62]

Amusement parks also quickly spread to foreign shores. In the first decade of the twentieth century an Englishman by the name of William Bean brought American park ideas to Pleasure Beach at Blackpool, a seaside resort fifty miles northwest of Manchester. Bean imported La Marcus Thompson's Scenic Railway (1907) and the Monitor and Merrimac battle show (1910). Annual trips back to the United States kept him up on the latest innovations.[63]

## Electro-Mechanical Saturnalia—For a Price

The amusement park was basically a commercialized, branded instrument for satisfying the saturnalian longings of festival goers and pleasure garden revelers—but with distinctively modern means and themes. Amusement parks brought pleasure because of their sheer concentration of multisensuality. In this conclusion let us focus on three ways that these saturnalian longings shaped modern values: the commercial exploitation of the pursuit of the new, the selling of a return to childhood, and the frantic commercialized quest for "fun."

The unrelenting vending of novelty was at the heart of the amusement park. Especially in its early years, the amusement park created an expectation of the ever new and improved, with each year offering new attractions, just as phonograph makers introduced new models almost annually. Here as in the packaging of other pleasures, the leitmotif was the creation of desires, which is one reason amusement parks featured fantasy structures and "novelties." The goal was to serve up the "wow" along with the popular and novel. Yet there were limits to this appeal. The inability to offer new rides and spectacles actually led to the downfall of Frederic Thompson at Luna Park, when he was forced out by his creditors in 1912. Novelty-based thrills brought in the crowds but could not keep them coming back because even a new ride, once experienced, was no longer new. To make the ride or spectacle worth repeating meant going beyond the "wow" of the unexpected or daring; and those who found success typically did so by evoking other emotions—controlled fear or vertigo, voyeuristic astonishment, nostalgic memory, or even sentimental patriotism—as Disney learned to do so masterfully in his theme parks later in the century.[64]

Successful theme parks also managed to create—and exploit—the longing of adults to return to some "fresh experience" recalled from youth. This did not necessarily mean sharing a moment of "wondrous innocence" with a child (as was done with the Kodak camera). Amusement parks were originally places where adult visitors sought liberation from adult duties—especially the burdens of family life. Children were certainly present at Coney, but photos show them mostly at the seashore, and not so much on the rides of Luna Park or even the seemingly childlike slides and revolving discs of Steeplechase. In photos from 1910, while kids appear splashing in the surf with their mothers, riders enjoying the toboggan slide were typically adults in suits and long dresses. Luna Park even offered nurseries for tired mothers. British writer E. V. Lucas, visiting Coney Island in 1921, was struck by the absence of children from New York's "safety-valve" (as he described the complex): "It is as though once again the child's birthday gifts had been appropriated by its elders; but as a matter of fact the Parks of Steeplechase and Luna were, I imagine, designed deliberately for adults."[65]

In the first decade of the 1900s, amusement parks were not cheap, and visitors were mostly young wage earners not yet burdened by the costs of raising kids. The absence of children suggests more, that adults free of parental responsibility were able to act like children—if only for the afternoon—leaving behind the restrained decorum of Victorian adulthood. This stepping back into boyhood and girlhood released psychological and physical tension, just as the traditional saturnalia had done—but in a manner far less threatening to authorities. The modern saturnalia turned (only) age upside down, leaving the rest of the world intact. Rollin Hartt in 1907 commented on how adult riders thrilled to the roller coaster: "the bright face of danger, challenging the eternal juvenile within you, seems—exactly as in years gone by—to be taunting, 'Fraidycat!' . . . You utter the cry of a tiny boy, 'Scare me again! Scare me—scare me worse!' "[66]

Frederic Thompson also understood how his Luna Park induced adults to regress: "People are just boys and girls grown tall. Elaborated

Photograph by Brown Bros., New York.

THE HUMAN TOBOGGAN GIVES THE CELLAR DOOR CARDS AND SPADES.

**Figure 7.6** The fun of grown-ups on a slide at Coney Island. Note also the men watching women enjoy that ride. *Everybody's Magazine*, July 1908, 29.

child's play is what they want on a holiday."[67] Edward Tilyou, son of the founder of Steeplechase Park, observed that because most people "look back on childhood as the happiest period of their lives," this was "the mental attitude they like to adopt" at the amusement park. And, when they did, they spent more. It is not surprising that young adults increasingly had fond memories of childhood: parents were becoming more indulgent of their offspring, and adults were coming to identify the childlike with self-indulgence. Men especially found escape from the Victorian world of hard work and competition—and providing for family—in the playful trip to the amusement park. Spending on "the child," even if only "the child within," had become a virtue to the middle class. Grown-up boys and girls could pay to play without ever violating still prevailing genteel norms because they were only being "children."[68]

What made this Peter Pan syndrome so appealing, though, was more than an escape from adult responsibility, and more even than a return to an imagined joy of childhood. It was a full-throated embrace of the modern idea that the individual has the right to earthly happiness, to have fun, not just the obligation to work, serve and achieve in hope of a better life to come. And, instead of the Victorian ideal of happiness as contentment, the amusement park offered the all-enveloping sensuality of the commercial pleasure site.[69] The crowds that flocked to Coney Island, Riverview, Cedar Point, and hundreds of other amusement parks sought stimulation of the imagination and senses. They shared with their ancestors many saturnalian delights and intrigues — social inversion, mockery, and a fascination with the supernatural and abnormal — delights and intrigues now powered by the engines and inventions of industry, with a fixed price of admission. The amusement park created a kind of electromechanical saturnalia, filled with monumental pleasure-making machines, offered to a public willing to pay. It was a sited package or really a cluster of pleasure packs with roots in some of humanity's oldest sensual quests, including festivals and pleasure gardens. But like most of the other packaged pleasures treated in this book, the amusement park was the product of a new era of compressed sensuality, a kind of electric theater (or mechanical circus) made possible by new techniques for packaging and commercializing pleasure.

# Pleasure on Speed and the Calibrated Life

*Fast Forwarding through the Last Century*

This book tells the story of the rise of the packaged pleasure, especially in the generation or two straddling 1900 in the United States, when mass marketing and rapidly advancing technology produced a unique cluster of commodities. We offer a particular slant on the broader rise of modern consumer society, the contours of which become clearer only as we fast-forward from 1920 to today.

Of course we should keep in mind that these commodities are part of an emerging consumer society, a "shopping culture" wherein people buy rather than make things for themselves, and consume a far greater range of goods than ever before. Thanks to mechanization and the rise of modern marketing, pleasures once rare, often social, and even free have become commodified, mass produced, and promoted for sale in personal portions and often consumed in private. What we buy has become part of our identity, shaping how we distinguish ourselves from and interact with others.[1] But consumer society has also transformed our sensory experience.

These sensual encounters are new in a Big History sense—and we need to realize that the long-term consequences are not yet entirely clear. Still, two broad processes seemed to be at work during the twentieth century and beyond. The first is *intensification*. Pleasure engineers have learned how to ratchet up sensory intensity to compete with satiety and boredom, helping them also to prevail over competing industries. Soda bottles have gotten bigger; pinball games have morphed into lightning-fast interactive video games; and movies have become more visually stunning. Even roller coaster engineers have learned to take the ride to the physical and psychological edge. Intensity begot more intensity as the novel and thrilling became common or even boring, requiring more novelty and thrills to attract consumers.

Intensification means more than packing more sensuality into the package, however. It means more art on the package and better access to it. New marketing techniques, including labeling, have made packaged pleasures ever more attractive and convenient to purchase and to use. The rise of the discipline of marketing is part of this, as anyone can see from how deeply trademark logos have dug into our minds. Who on the planet will not recognize images of Marlboro or Coca-Cola? Thanks to technology, consumption has also become easier and often more "immediate": many goods (and bads) can now be purchased anonymously online, dispensing with the intermediaries of clerks or soda jerks. Immediacy and convenience of access were early on keys to the attraction of phonographs, cigarettes, and canned foods; and today we see a similar allure in the automatic shuffle of thousands of tunes on iPods or streaming audio players. There is scarcely anything today that isn't sold in single-use portions, portions that very often pack an ever-greater punch.[2]

Much of this is the consequence of new technology—combined with aggressive, science-based marketing. The mechanical and electrical innovations of the nineteenth century were transformed by the development of electronic, laser, and especially digital technology in the twentieth century. Reproduced sounds became (first) louder, more faithful, and eventually hyperauthentic; images appeared in ever-higher resolution, in some instances flashing faster, and now even in 3D. Personal and now portable computers have eliminated much of the wait for and scarcity of pleasure, allowing bits and hits of sound and sight once provided by records, snapshots, and magazines to be delivered in an uninterrupted flow of sensation that would have shocked Edison and Eastman. Computer-assisted engineering and linear synchronous motors have also taken the thrill ride to the (apparent) edge of human pleasurable endurance. The packaged pleasure has sped up personal life, making experience less dependent upon the slow pace of nature and the inconvenience of other people's annoying ways and delays, while also liberating us from many of life's boring bits. And so accustomed to this have we become that we may fret over a few seconds' wait for any of the now converged inputs of music, images, movies, video games, mail, and text, or voice on our smart phones. For some people, these products truncate the traditional delights of "anticipation" and even memory.[3]

But intensification is not the only longer-term trend in the history of the packaged pleasure. A second is its seeming opposite, *optimization*. Merchandisers have recognized that more subtle sensual encounters often meet with greater success than the steady escalation of sensation. There are obvious physical, psychological, and cultural limits to intensification; and pleasure engineers have learned to look for ways to optimize deliveries to maximize sales. Thanks to the unrelenting quest of marketers to colonize space and time, consumers have learned to desire candy bars and soda, even popular songs and TV sitcoms, as essential complements to the routine rhythms of the day, week, or season. In contrast perhaps to the intensified package, the optimized pleasure, measured to fit into the flow of daily life, seemed to be no threat to the work ethic (often the opposite) or to physical or mental health (at least in the short run). These delivered pleasures became a second nature to many, key to the modern calibrated life.

So in this penultimate chapter, let us fast-forward through the twentieth century and into the twenty-first to see what happened to the innovations discussed in this book. We will, as it were, travel in H. G. Wells's "time machine," stopping briefly and selectively to illustrate how the packaged pleasures that emerged in the decades around 1900 have come to envelope and define our lives today.

## Supersized Soda and Overflowing Shopping Carts

Packaging has the prosaic but still astonishing ability to contain stuff that would otherwise decay, dissipate, or disappear, while also advertising contents to consumers far and wide. In the short space of just over a century, this double role of packaging gradually reached almost every corner of human experience. Consider again the modest case of sugared fizzy drinks.

Bottled soda began as something sold and consumed in restaurants or drugstore fountains, but through the packaged pleasures revolution became widely available to the person on the go. By the early 1920s it was widely sold in vending machines and for home use in six-bottle "homepacks," bought at newly emerging supermarkets. Coin-operated vending and domestic consumption reaffirmed the same kind of unmediated access and private enjoyment already set in motion by the gumball machine and the phonograph. The process advanced apace: from 1913, motor-powered trucks delivered soda over a much wider territory than was ever possible by horse-drawn wagons, and metal coolers introduced to stores in 1924 offered chilled refreshment at will (a process advanced with the home refrigerator, introduced in 1916 and widely disseminated only a generation later). The cumbersome glass bottle began its long decline in 1936 with the invention of the soda can, first awkwardly shaped like a bottle with a neck and cap, but from 1940 on as a flat-top cylinder. Lightweight aluminum cans came in 1957, followed by can-dispensing vending machines in 1965. In 1970, plastic bottles replaced much heavier and more easily broken glass bottles that had also been slower to cool. By 1965 nonreturnable bottles challenged the expensive and inconvenient (but eco-friendlier) consumer routine of paying a deposit for and then returning used bottles.

All this made shipping soda cheaper and drinking it more convenient. The never-ending quest for convenience hit stride with self-contained can opening, first with the ring pull tab (first used for beer in 1962) and then the less troublesome stay-on-tab in 1974. More-to-drink-packages went along with easier-to-drink-packages. In 1936, Pepsi Cola cut into Coke's dominance by offering a twelve-ounce bottle for the same price (5¢) as Coke's six-ounce bottle, the first of many supersizings. By 1960, annual per capita consumption of soda pop in the United States had risen to 185 bottles and by 1975 would reach 485.[4]

More important still for intensification was the substitution of high-fructose corn syrup for sugar from the late 1970s. Unlike sucrose, this cheap sweetener does not produce insulin secretions (that make people feel full), but unfortunately, it does stimulate fat cells. The success of the seemingly innocuous soda pop has played no small role in the increase in average daily caloric intake by Americans—from 3,300 to 3,800 over the last three decades of the twentieth century. This trend in drinking empty calories (and often caffeine) coincided with the decline in milk consumption—and a dramatic increase in childhood obesity.[5]

Other upgrades in packing foods were more benign. Flash freezing allowed the shipping of otherwise fast-spoiling seafood over long distances and offered a better-tasting way of preserving peas, corn, and soft fruit than in cans. Though the technique was developed in the 1920s by Clarence Birdseye and sold to General Foods in 1929, fast-frozen foods were rare in American kitchens until after World War II—because of the expense and thus rarity of retail and domestic freezers. The United States Air Force introduced frozen individual meals on overseas flights during World War II and, in 1947, Carl Swanson's frozen turkey business exploited this innovation, developing what became known as the TV dinner. If not noted for fine cuisine, Swanson at least provided nearly effortless home dinners and, more important, the individualized meal reinforcing a retreat from the shared bowl.[6]

Freezing and other modern packaging techniques required centralized manufacturing and distribution but also new retail outlets, especially the chain grocery store and the supermarket. The neighborhood grocery store with its barrels of flour and walls of shelves containing jars of generic foodstuffs behind a sales counter was already on its way out in the late nineteenth century with the rise of the chain grocery and variety store. By 1870, the Great Atlantic and Pacific Tea Company (later shortened to A&P) of New York distributed tea and coffee to an expanding string of small retail outlets, adding canned and boxed goods in 1912. A&P had about fourteen thousand stores by 1925, with customers benefiting from a new type of store layout, invented about a decade earlier. In 1916, Clarence Saunders of Memphis had introduced self-service with Piggly Wiggly, a new type of grocery store with checkout aisles and open shelving, eliminating the counter between the customer and the goods that required a clerk's assistance. In the 1920s, Southern Californians took this one step further by inventing what we now might regard as the familiar modern supermarket (Ralph's, for example), complete with parking lots and the promise of one-stop shopping. Other retail innovations took place on the East Coast, as, for example, in 1930, when Michael Cullen opened his first bare-bones warehouse-style grocery store in the New York City borough of Queens with the promise of rock-bottom prices for Depression-stressed shoppers, anticipating big box stores like Costco, Walmart, and Sam's Club.[7]

Many of the foods sold in these new supermarkets were essentially snacks, quick-energy foods one could eat anytime anywhere outside of traditional meal times. Of course, candy and ice cream already had a long history in urban areas, and cakes and other "goodies" had been common delights at seasonal fairs. But the variety of and access to snacks expanded dramatically in the second half of the twentieth century—and not just confections, but myriad other kinds of fat- and salt-laced treats. "Street foods" were already being denounced by nineteenth-century dietary authorities as a threat to the family meal, but many families still snacked on pretzels and other salty treats at special events like county fairs. Yet, in the 1890s, when Cracker Jack was sold at retail stores as an impulse purchase, we see an early step toward the routine snack treat. Improvements in cellophane and other plastic packaging in the 1930s led to an explosion of fatty snack foods. Potato chips were available in the late nineteenth century, but were not terribly popular because they were sold out of barrels and quickly became soft and soggy. The new plastic wrapping was exploited by Herman Lay of Atlanta, who founded his famous brand of Lay's potato chips in 1937. Partnering with Elmer Dolin, the maker of Frito Corn Chips, Lay's and Frito merged in 1959, offering a continuously expanding array of salty snacks (the whole of which was then sold to Pepsi in 1965). The cumulative trend was anytime snacking (aided by personalized packages), easily accessible and often consumed during personal or unoccupied time like driving a car, reading, or watching TV. Packaged snacks

**Figure 8.1** A photo by Clifford Saunders of the innovative design of the Piggly Wiggly self-service grocery store in Memphis, Tennessee, in 1918. Note the turnstile and structured layout to direct the customer to pass by all the goods. Library of Congress (LC–USZ62–91202).

supplemented and sometimes even replaced the effort and sociability of the prepared meal.[8]

One quintessential food package was the fast-food meal, especially the hamburger. The ground beef patty, introduced in German-American restaurants in the 1870s, was sold on a bun by American city street vendors from the 1890s. By 1921, the hamburger stand had morphed into a retail chain known as White Castle. These ultraclean, boxlike restaurants, containing a grill, counter, and white-clad hamburger flippers, were the invention of Wichita hamburger-stand owner J. Walter Anderson. The White Castle chain spurred many imitators, often with "white" in their names. In 1940, an even more momentous shift came when Richard and Maurice McDonald opened a hamburger stand in San Bernardino, California (60 miles east of Los Angeles). By 1948, the brothers had revolutionized the drive-in restaurant with their reduced menu and assembly-line preparation, offering bagged food purchased at a window for consumption in the car. All this proved to be more profitable than the old-style service with individually prepared orders on plates with cutlery, delivered to table (or car) by waiters. The concept was quickly copied by Burger King (1952), Kentucky Fried Chicken (1952), Carl's Jr. (1956) and Taco Bell (1962), with each becoming a multibillion-dollar franchise chain. McDonald's was sold to Ray Kroc in 1954, leading to a new, tightly controlled franchise system. Indoor seating would return to the franchise drive-in restaurant in the 1960s, but in 1969, Wendy's reintroduced in-car eating with the drive-through window. The merger of fast food with the mobility and privacy of the car was seemingly unstoppable.[9]

As fast-food restaurants accommodated the time constraints of the two-income family, the convenience of the drive-through also increased demand for faster home cooking. Convenient preprepared foods like cake mixes saved the time of gathering and mixing ingredients from scratch. But more recently, even these have begun to disappear from shopping carts, as fewer and fewer people are willing to spend even thirty minutes waiting for a baked cake. Today, many boxed or frozen meals are still on the market only because the microwave has radically reduced cooking times.

Of course there has always been resistance to sugared and fatty snacks and fast food. There is a long history of opposition to packaged fare, and vociferous movements to return to natural and organic forms of eating date from the nineteenth century (see more on this in chapter 9). One commodified expression of distrust of super-sugars was the push to produce low and eventually no-cal sodas, introduced nationally with Diet Rite (a cola) in 1958. This was followed quickly by other low-cal drinks like Tab, Diet Coke, and many others sweetened by diverse sugar substitutes. Low-fat processed foods have filled store shelves since the 1960s, offering the taste and comfort of fat, sugar, and salt, presumably without the consequences of weight gain and other health-threatening conditions. And, of course, news of how cigarettes caused cancer and other diseases prompted tobacco companies to introduce "filtered" cigarettes in the early 1950s—with the false promise of a "safer smoke" (false because smokers inhaled just as much smoke that was no less deadly). Tobacco companies also manipulated the chemistry and physical design of cigarettes, allowing smokers to extract however much smoke they needed to achieve "satisfaction."

More subtle were efforts of packagers to optimize the pleasure dose in tobacco and superfoods. This took many forms, most of which were designed with the goal of increasing consumption and therefore profits. The eternal quest for ever-milder cigarettes, begun with flue curing, flavoring, and eventually mentholation and other tricks, encouraged youth initiation and chain smoking. Consumed throughout the day, these relatively mild cigarettes served as mood stabilizers for millions, "lifting" them when tired or bored, calming them when nervous or stressed (from not having gotten their "fix" of nicotine), again a consequence of optimization.

Less deadly was the discovery by candy and later snack food makers that the simple "jolt" of sugar and fat could be augmented by adding ever more complex flavor "profiles." The carefully designed and complexly layered Snickers bar is one classic example of optimization. Eaters biting into the semisweet chocolate shell encounter the contrasting tastes of supersweet caramel with salty peanuts, followed by the smooth nougat reward of the candy bar's core. This was not just a matter of appealing to ever more sophisticated palates or moving beyond the hard candies once given to children. David Kessler and others have shown how the food industry has learned the arts of "loading and layering": slathering potato chips with coats of (real or artificial) cheese, spice, or sour cream, for example. The net effect has been to increase the fat, sugar, and salt content of even formerly simple foods like crackers and salads (with calorie-packed "dressings"), while simultaneously creating novel sediments of flavors. The goal of these new foods is not so much to overwhelm or intensify sensation as to optimize flavor mixes—by enhancing "mouth feel," for example. Fat is also used as a lubricant and vehicle for carrying flavors, just as sugar or caffeine are used as stimulants. The "dynamic contrast" of flavors and textures so successful with the Snickers bar has become common in a wide range of processed pseudofoods, revving up the attraction. The key is not just larger portions or even more complex flavors, but rather a balance in which the mix of salt, fat, sugar, and flavors are optimized for appeal (i.e., sale). Food companies sometimes even mask their incorporation of sugar or salt into foods (like ketchup or salad dressing) by giving the sweetener a new name—fructose, brown sugar, molasses, high-fructose corn syrup (now "corn sugar"), and so forth. Sugar and saturated fats are also used to extend the shelf life of foods, and other modifications are designed to ease their transport. Food makers are thereby able to make eating more appealing but also less nutritious and, in ways not previously possible, more habit forming.[10]

## Packaged Sound Everywhere, Anytime

The last century has brought a similar pattern of increasing access and intensity (along with integration into life routines) in sound recordings. Until the mid-1920s, the recording of sound was based on the simplest of technologies: the direct mechanical reproduction of sound via the spiral grooves of shellac records. Edison had hoped to use electricity to transfer sound from the studio into the home record player, but that would occur only after the appearance of a new and more sophisticated technology: the radio, which brought with it technologies essential for electric recording. Wireless transmission required a relatively complex process of converting sound from a live source or acoustic recording to an electric signal, which could then be broadcast via electromagnetic waves to a receiver that would turn that signal back into sound.

The key to radio was the discovery in 1863 that electrical circuits could radiate electromagnetism without the wire required in the telegraph. Harnessing this wireless energy as a signal came when Heinrich Hertz invented a spark gap transmitter of radio waves in 1888. A few years later, Guglielmo Marconi turned this into a practical device, the wireless telegraph, based on a system similar to the wired telegraph, involving the sending and receiving of electromagnetic (rather than electric) pulses that could be read as code. By 1906, Reginald Fessenden had reproduced the telephone's capacity to transmit voice and music with his arc or high-frequency generator, creating radio waves that varied in their width—converting sound into amplitude modulation, or AM radio signals. This combined with improvements in reception to make the modern radio. In 1904, John A. Fleming developed the two-electrode vacuum tube that served as a valve to "rectify" alternating current radio waves (by moving the electrical signal in one direction as pulsing direct current, which was convertible into sound waves for listening). Two years later, Lee De Forest completed the revolution by adding a wire grid that amplified the signal, vastly improving the volume and range of reception. In the next decade, however, the radio still served mostly as a wireless telephone for business and military communications. It was not until 1916 that David Sarnoff, future chief of NBC, proposed that the radio receiver could also be an alternative to the home phonograph, a shift analogous to the Internet's recent role in supplanting CD recordings by downloadable audio files. Record and phonograph companies in the 1920s hoped that this potentially devastating threat from radio would be delayed because of patent conflicts between major players in radio technology, especially AT&T, General Electric, and Westinghouse. The U.S. government encouraged a patent pool during World War I, however, and this led to the consolidation of this complex technology in the Radio Corporation of America (RCA).[11]

And so it was that AT&T's Bell Laboratories, not Victor or some other record maker, developed electromagnetic microphones for use in broadcasting radio — rather than for recording. A quarter of a million radio sets were in American homes by 1921, with service provided by a burgeoning array of stations. Many of these joined the National Broadcasting Company (which had links to RCA) in 1926, and others affiliated themselves shortly thereafter with the Columbia Broadcasting System. These networks delivered programs nationally via AT&T's long-distance telephone lines to local radio stations for broadcasting to home radios, producing a wealth of "free" music and entertainment, all financed by network advertising (especially of cigarettes and soda and the like). Combining radio tube amplification with the invention in 1925 of the modern loud speaker, the radio could be enjoyed at high volume with no need for earphones. All this led to the near collapse of the recording and phonograph industry by the late 1920s.[12]

But radio also helped to transform recorded sound by forcing phonograph makers to abandon the horn and diaphragm of acoustic

**Figure 8.2** An astonishing range of promises came with the radio, including bringing "color" into the dull routine of personal life. *Saturday Evening Post*, February 22, 1930, 88.

recording for the amplifier, electromagnetic microphone, and loud-speaker developed first primarily for radio. Completing the switch to electric phonography was the electromagnetic cartridge and stylus in 1925 for the playback of records through amplifiers and speakers. Victor in 1925 began making electric records playable on a special acoustic phonograph capable of handling the wider frequency range afforded by electric microphones. By 1928, the company had wired its record player (with the new cartridge) to the RCA radio amplifier and loudspeaker, introducing an all-electric phonograph system. Die-hards, especially Edison, were reluctant to abandon the "authenticity" of acoustic recording and playback. But no one could deny that the electromagnetic microphone increased the volume and musical range by two and a half octaves over the mechanical diaphragm, producing an aural experience far closer to what was heard in a concert hall. In time, sound engineers would manipulate the volume and tone gathered from multiple microphones to record a sound that was "superior" to the live concert experience. These innovations dramatically intensified the sensation of recorded sound.[13]

As it turned out, the new record and phonograph did not immediately fare well in the market, given the convenience and low cost of radio receivers. Broadcast radio required no special effort to listen, apart from turning on and tuning in. Radio was also free after the initial cost of the receiver—and even this expense dropped sharply in the 1930s with the introduction of cheap compact radios. There was, of course, less choice on broadcast radio than in the record catalogs (one could only change the channel), but even middle-brow programs like Paul Whiteman's *Kraft Music Hall* offered variety. Radio was a big threat to record makers, and in 1929 the financially troubled Victor was sold to RCA, while Edison stopped making records and phono-graphs. By 1932, record sales had dropped by 94 percent from levels only five years earlier. New companies like Decca drastically cut retail prices (from $1 or more per disc to only 35¢ or even 25¢) and Colum-bia tried to expand its appeal by recording the big band swing music of Duke Ellington, Count Basie, and Benny Goodman. Still, the De-pression took such a hit on personal purchasing power that fully half the discs sold in 1936 went into the 225,000 jukeboxes in bars, cafes, and fountains, an interesting regression back to the first decade of the phonograph when coin-operated models prevailed in public arcades.[14]

The private phonograph would return to glory only when longer recordings and higher sound quality became possible (and the economy recovered). Part of the problem was how little music each disc delivered. The standard four-minute shellac record was bulky and heavy, especially for what listeners got—a single song or short segment from an opera, chamber piece, or orchestral composition. The idea of a longer-playing record had been germinating for quite some time. From 1928, radio programmers produced electric transcription records to distribute shows to local stations (instead of depending on the live programming of the networks) and these had to be about 15 minutes long. Eventually, these transcriptions were made on 16-inch records (instead of the standard 10-inch disc), turning at 33⅓ revolutions per minute (rpm) instead of the standard 78 rpm, yielding about 15 minutes of entertainment between commercials. By the late '40s, the radio networks used these transcriptions to facilitate live programming broadcast simultaneously in all time zones without repeat performances.

But the breakthrough for consumers really only came in 1948, when Columbia/CBS's Peter Goldmark developed the modern long-playing (LP) record. Goldmark did two things. First, he replaced shellac (in short supply from the Second World War) with vinyl. This meant a lighter, better-sounding record (shellac had always produced a certain "hiss") and allowed for a microgroove process that could put far more music on a single record. Second, Goldmark copied the transcription discs, reducing the record speed from 78 to 33⅓ rpm, finally making possible the unfettered play of a classical music movement of twenty minutes. This led to a golden age of recording—and not just of classical music but soundtracks of whole musical shows. The first LP record to sell 1 million copies was the soundtrack from *Oklahoma!*, Rogers and Hammerstein's famous Broadway musical released in 1949, while another show by that same duo, *South Pacific* (1958) became the biggest LP hit of the '50s. In response to Columbia's LP format, RCA offered a smaller light-weight 45 rpm record in 1949, along with a specially designed fast-acting record changer (to accommodate the record's large central hole), allowing a rapid succession of 45's to be played. The

LP soon won over devotees of classical and other long-playing music, while the 45 came to dominate the market for short pop tunes, especially country and rock music, appealing to the mixed and varied taste of youth who had neither the money nor patience for the long-playing record.[15]

This division between the 45 and the LP paralleled the chasm between Victor's Black Label and Red Seal records a half century earlier. While teenagers played their stacks of pop tunes on their 45 changers, adults and self-assumed sophisticates listened to their Beethoven or even Frank Sinatra albums at twenty-minute stretches on their turntables. The difference was not only in the duration (and required attention span) but also in sound quality. The cheap 45 offered variety, ease of access, and durability, but not aural perfection. In this way it was similar to the AM radios that played disc jockey-introduced rock music from 1954. Teens cared more about timely tunes than rich and accurate reproduction—like those downscale consumers of Black Label Victor records in 1905. The contrast with the LP couldn't have been greater. From their very beginning, LP records by Columbia and others promised fidelity to the original performance, just as Red Seal records had done fifty years earlier. And the LP came far closer to doing just that, attracting the same audience that listened to the higher-quality FM radio then challenging AM. Even before the LP, new companies introduced high-quality audio receivers that met the demands of an emerging community of audiophiles. Garrard and Fisher offered component phonograph equipment in 1945, and magnetic tape recordings (and recorder/players) became available for home use soon thereafter (though for many years, tape recorders were used mostly in recording studios).[16]

The LP's packaged sound followed a familiar track: more sound, increased ease of access to sustained listening, and greater quality or intensity of experience. This last-mentioned point was reinforced with stereophonic recording. Already in the 1930s, recording studios were experimenting with cutting two separate tracks running on both sides of a forty-five-degree angle record groove. When the two tracks were recorded from microphones in widely separated positions on the sound stage, the result was two slightly different sounds. And when the two tracks were reproduced in two speakers (set similarly wide apart) in the home, the result was astonishing. The listener heard not what seemed to come from out of a hole in the wall (as from the old monaural recordings), but rather an "arc of sound" that made listeners feel that they were actually at a concert, with the best seat in the house (or so advertisers tried to convince audiophiles). Stereo records required special equipment, including a new needle and cartridge to pick up both tracks, and a new amplifier and speakers to channel the separate sounds, but the new technology was an instant success after its introduction in 1958. Listeners took great care in positioning speakers to get a sound that was imposing, making the living or even bedroom into a concert hall. Stereo became nearly a cult-like obsession, especially for male techies who purchased and often personally assembled or linked a cluster of audio components—including powerful amplifiers, tuners, cartridges, tone arms, turntables, and large and heavy speaker ensembles. The phonograph was also no longer disguised as a piece of furniture, as it had often been since the days of the Victrola; it was now more of a modernistic fashion statement, especially when festooned with buttons and dials that only the cognoscenti fully understood. Even more, these electronic wonder boxes delivered a stereo experience that, for some devotees at least, became an intensely focused activity, often to the exclusion of other people and cares.[17]

Innovations followed, advancing the appeals of the stereo enthusiast while also forwarding the promise of portability and easy access to an unlimited assortment of audio "hits." New recording formats also eliminated the annoying scratches and the limited twenty-minute playing time of the LP, while also attracting the same youth audience that had bought 45's in the 1950s—notably the cassette tape and CD, followed eventually by the mp3. This avalanche of innovations finally eliminated the phonograph record after a century of success. Very briefly this final trajectory can be traced—beginning with the transistor.

Since the 1930s, researchers at Bell Laboratories had been trying to find a solid-state replacement for the fragile, breakable, and relatively large electronic vacuum tubes that formed the core of the radio. Required was a semiconductor, a substance that could replicate the controlling and amplifying capability of the tube. The key breakthrough came in 1948, when Bell Labs scientists introduced the transistor based

**Figure 8.3** Though the 45-rpm phonograph was the domain of the teen in the 1950s, manufacturers targeted youth especially for portable LP phonographs. *Seventeen*, September 1959, 195.

on the semiconductor silicon, making possible smaller, potentially portable sound machines. Groups of transistors were miniaturized in the integrated circuit in 1958 and still more by the microprocessor chip in 1971. The transistor and its successors were small and required very little power, meaning that that they could be run on batteries for long periods. They also generated much less heat and only rarely burnt out. In 1954, Sony in Japan and later other manufacturers purchased licenses to develop the transistor's potential in portable radios. Hand-held transistor radios couldn't offer the sound quality of home stereos but they were portable, an appealing alternative to the cumbersome vinyl record with its requisite stylus and turntable. The cassette tape player, sold first in 1964, also did what no record player could do: provide music on-demand in the car or in the pocket. By the 1970s, some teenagers had begun replacing pocket transistor radios with the boom box, a larger, louder, big-bass combo radio and cassette player. A privatizing element was then introduced in the 1979, when Sony developed an inexpensive, pocket-sized, portable cassette player. Its most famous commercial use was the Walkman, a retro/revolutionary device that resurrected the earphones of early radios while simultaneously mobilizing (and isolating) the listener. The cassette also revived another old idea—recordability—that the phonograph had abandoned in the late 1890s. This had interesting implications in the realm of intellectual property, since the personal duplication of music also allowed ordinary consumers to circumvent copyright.

The cassette era, however, was relatively short-lived. In 1982, Phillips began marketing the laser-read digital compact disc, or CD. Made of clear PVC plastic, the CD was coated with a thin layer of aluminum onto which digital code was imprinted. The CD's digital reproduction did not copy the sound wave of the performer as did analog tapes and records; instead, it "merely" sampled that wave (about fifty thousand times per second) and encoded these samples as a series of pulses or digital "bits." Few listeners, though, could tell the difference. Even more, the CD extended once again the duration of play, allowing eighty minutes of music, the equivalent of both sides of an LP without the bother of having to turn it over. And the CD track did not wear the way that the record groove did—and didn't have the distracting background noise of the cassette. This appealed to both the old devotees of stereo LPs and the young hooked on 45's. Even more, digital duplication was far faster than analog tape reproduction. Within six years — by 1988 — CDs were outselling LPs and threatening even the cassette. And then in 1991, stores abruptly abandoned both vinyl and tape.

The digital CD victory, though, was short-lived. By 1996 sales had stalled and a radically new digital technology was starting to dominate, based on the personal computer and the Internet. The ease of copying digital recordings, especially in the compact mp3 format, allowed hundreds of hours of music to be stored and accessed instantly on digital devices, including personal computers. While the era of stereo and hi-fi had stressed sound quality, now most young listeners, eager to adopt the latest technology, were willing to accept "decent sound" as long as the price was right and musical selections were vast and unfettered. Seemingly unlimited freedom of choice became the engine powering this latest music revolution, and with this Internet-driven revolution came a shift from corporate-controlled delivery of music to an "endlessly shifting menu" of (often) eclectic tastes.[18]

The intensification of packaged sound, then, yielded two important forms of musical fruit: concert fidelity and immediate access, both of which also made it possible to listen to music in private. Over the course of the last half a century, this has meant some decline of live music. Yet it has recently flourished in bars with local musicians and in the concert tours of recording artists with CDs even used to promote these live shows that offer a unique stage performance and crowd experience. Music, like so much else, has become stockpiled in accessible packages for personal use whenever and however.[19]

The impact of all this is certainly more ambiguous than the packaged use of tobacco or even the substitution of junk food for the planned meal. But the new formats of packaged sound have shaped social and sensual experience. Sociability takes many forms, of course, and along with the personalization and privatization of the listening experience have come new forms of sociability as listeners construct and compare playlists in CD or mp3 form. Yet the personalization of the listening experience has been accompanied by a heightened aural focus, with earphones being used more often to block out sounds than to share with others. Once again, the packaged pleasure has been intensified. Recorded sound has also become part of an ongoing background, a kind of mood adjuster used to block out less attractive parts of the world.

Long ago, in fact, Edison had understood the potential of using recorded music as a means of mood adjustment. Beginning in 1934, the Muzak Corporation distributed canned music to hotels and restaurants and then to business subscribers. In the latter case, the idea was to maintain the productivity of factory and office workers by varying the music according to the presumed energy needs and moods of the day (upbeat rhythmic music to fight sleepiness at lulls in the day like 10:30 a.m. and 3:00 p.m., for example). It was "non-entertainment" for people to "hear, not listen to," according to a Muzak executive in the 1950s. Since the beginning of the twenty-first century, iPods have become a kind of Muzak self-administered to millions, with the sound kept flowing and interruptions from conversation or the sounds of nature kept to a minimum.[20]

## Pictures at a Quickening Pace

Apart from the shift to color film in the 1960s, the image captured on film changed relatively little for nearly a century after the invention of celluloid roll cameras in 1888. Yes, the Polaroid camera (1949) offered instant photography, replacing the centralized developing process with a few seconds of self-developing on special Polaroid film. But the quality was never to equal older commercial processing, and the partnering of the amateur photographer and centralized corporate processor survived—until the digital revolution.

Just as the digital revolution transformed packaged music, so too did it dramatically impact the snapshot. The key change was in substituting the chemical process of capturing images on film with an electronic process involving the fixing of an image via tiny sensors—charge-coupled devices (CCDs)—invented by William Boyle and George Smith at Bell Labs in 1969. CCDs measure light intensities in tiny discrete points, transforming these "pixels" into electrical charges. When transferred to a computer file row by row, these pixels can be stored as digital code. As grids of pixels, digital images can be reproduced like other digital files. And since CCDs are small and arranged in a dense grid, like the audio CD they can "sample" a vast number of points of light, creating high-resolution images. CCDs made it possible to scan, copy, store, and transfer images electronically. And though this technology was first quite bulky and expensive and used only in fields such as astronomy, medicine, and the military, advances in data compression and storage (and lowered costs of manufacturing) allowed a personal digital camera to hit the market in 1988. Digital cameras eliminated the wait and waste of conventional film development, but they also made it possible to take hundreds and eventually thousands of pictures without "reloading." Kodak's old promise of plenty in the twelve- or twenty-four-exposure film roll was enlarged by orders of magnitude. The sheer quantity of images had obvious appeal, but has arguably also worked to diminish the value of the singular or unique image, just as the Internet/mp3 revolution made the special song less special.[21]

A related trend can be seen in the quickening pace of images in movies. The "cinema of attractions" launched the modern fascination with moving images, dazzling viewers with what could not be seen on the stage or in ordinary life. As movies became more about narrative, the isolated visual thrill became less and less central. By the mid-1920s the "soundtrack" was assisting in this transition to narrative film, one of the most profound examples of this new ability to transform sound into electricity and back again. Several different technologies competed for dominance in these very first "talkies," including separate synchronized electric disc recordings played back to match the action of the film (as in *The Jazz Singer* of 1927). But the sound-on-film method ultimately prevailed. In this instance sound waves, converted into electrical impulses, modulated the intensity of an electric light, the varying patterns of which were then photographed and copied onto the running edge of the movie film. On the projector, light shined through the sound track recreated these varying intensities of light, which were converted via a photoelectric cell to modulating electrical current run by means of an amplifier—and then converted back into sound via an electromagnetic loudspeaker.[22]

The coming of the talkies is a familiar story, another chapter in the history of packaging pleasure. Music and motion were wedded in an amazing era of movie musicals and song and dance extravaganzas, with some of the most notable being those produced by Busby Berkeley in Hollywood in the 1930s. The soundtrack made the words and voice of actors come alive for audiences, deepening also the cult of the celebrity star—as will be known to anyone who has seen the black-and-white flicks from this era. But the studios were actually slow to adopt some of the new visual technologies—notably color film. Dialogue, rather than special effects, prevailed in the melodramas and comedies that dominated the golden age of Hollywood. Even the introduction of 3D movies and widescreen Cinemascope in the 1950s did not displace the centrality of story and dialogue.

Many films changed in the 1970s and 1980s, however, with the quickening pace of the moving image. Not only in movies and video games but in TV programs and especially commercials, the duration of the photographed shot decreased as efforts were made to increase visual intensity. This was driven, initially at least, less by technological breakthroughs than by cultural changes, notably the replacement of older moralistic narratives (especially in westerns and crime movies) by more action-oriented scripts.[23] Think of Clint Eastwood's role in *Dirty Harry* (1971) and its sequels. This is about more than an angry San Francisco cop (Harry Callaghan) fighting police bureaucracy in his quest to bring a serial killer to "personal justice." The story is driven by timed "beats" of brutal action throughout the film, with relatively little (if any) character development on Harry's part.[24] Violent action was further aestheticized in the quasi-mute martial arts movies of Bruce Lee, Jackie Chan, and others with flying fists and feet. By the mid-1970s, the action-filled heroics of Sylvester Stallone and Arnold Schwarzenegger had gained popularity, moving the fast-paced clash of good guys and bad from Hollywood's B- to A-list of filmmakers.[25]

As narrative atrophied, at least in adventure films (if not in dramas or comedies designed for older audiences), the pace of action noticeably accelerated. Consider the fact that the number of "kills" increased from 15 to 162 from Bruce Willis's first *Die Hard* (1988) to its second sequel 7 years later, *Die Hard with a Vengeance*. It is not just that the "beats" came faster. The story was sacrificed for an aestheticized flow of mostly violent action.[26]

Fast-cut action films rose in tandem with the digital video game, closely following the miniaturization of the computer. William Higinbotham's 1958 invention of a game called *Tennis for Two* (later *Pong*) was the humble start. Essentially a digital dot bouncing off dashes, this first video game was little more than a diversion for computer scientists, followed in 1961, with *Spacewar!*, in which primitive spaceships annihilated one another with blips of digital light across a black and white screen.[27] It took another decade, however, for the video game to be packaged for arcade and home use. Nolan Bushnell's electronic Atari arcade games (1972) were pivotal here, supplanting the mechanical/electrical pinball games dating from the 1930s. In a pattern familiar from the home use of the phonograph, Atari offered a home unit for *Pong* (in 1975), and in 1977 the Atari 2600 allowed a variety of game cartridges to be played through TV sets. More exciting and fast-paced games appeared with *Space Invaders* (1978) and an increasing number of Japanese innovations, including the wildly successful *Pac-Man* (1980) and *Donkey Kong* (1981). Sharp competition and the flooding of the market with look-alike games (since game cartridges from many different companies could be played on Atari's and other consoles) led to a temporary collapse of the video game craze in 1983. But the Nintendo Entertainment System resurrected the thrill two years later, with a more powerful player and strict controls on which company's cartridges could access the system. Nintendo's digital hero Mario evaded barrels and other trash thrown at him by a gorilla; Mario, however, was soon challenged by Sega's Sonic the Hedgehog, who whizzed through tubes and over digital obstacles.[28]

As the visual palette grew richer with larger game files in the 1990s, the pace of the action increased, especially in the form of graphic violence. Early examples are *Street Fighter* and *Mortal Kombat*, first appearing in arcades and bars and then the home. The intensity of kill-or-be-killed games escalated rapidly with *Doom* (1993), a sci-fi/horror first person shooter game marketed by id Software.[29] Driving this was the shift of sales beyond children to young adults, whose insatiable demand for ever-faster-paced games led to hits like *Grand Theft Auto* (1997) and *Halo* (2001).[30]

Again, like so many other packaged pleasures, the video game has done more than ratchet up the quantity and intensity of sensual experience. Manufacturers have tried to immerse players in the flow of the game, creating a deep, sometimes even addictive, engagement of players. The movement of joysticks, buttons, and mouse that animated the once-thrilling games of *Pong* or *Pac-Man* and so many that followed was transformed by Nintendo's Wii game system (2006), where the controller tracks the player's body movements, creating a more perfect illusion of being in the game.[31] Game designers have added new layers of image, sound, and tactile sensitivity, encouraging players to play "for life," and not just during childhood. Game makers alternate doses of challenge and payoff to entice the action, and psychologists have even been hired to identify the frustrating and boring parts. So powerful are these games psychologically and even physiologically that many players who grew up on Atari or Nintendo did not abandoned these "toys" once they grew up; a surprising number continue to play ever more intense games deep into adult life, sacrificing time for relationships with family and friends, or time spent on some lasting and less "virtual" form of self-improvement.[32]

The quickened pace of the flashing image is also a phenomenon seen in TV shows and ads. The widespread use of the TV remote control in the 1970s (invented in 1956, then a clumsy wired device) and the proliferation of channels with the introduction of cable and the VCR in the 1980s, led directors to introduce ever-quicker camera shots in hopes of attracting and holding audiences. Armed with an itchy trigger finger on the channel changer, viewers have learned to demand instant and continual entertainment. Camera shots have been made to jump around within a scene but also have become "quickly interleaved from different angles to show multiple split-second views of the same scene—'double cutting' or 'triple cutting,'" as James Gleick writes. MTV set part of this trend in 1981, with its broadcast of fast-paced music videos. More recently a competing cable channel, E! Entertainment, introduced shows consisting of little more than "a pastiche of clips, interviews, promotional tapes, and similar fare, all designed to be glittering enough to hold the attention of channel surfers whenever they happen to drop in."[33]

The pace and length of TV ads has followed a similar path. In 1978,

**Figure 8.4** A late-1970s video-game arcade, paralleling earlier arcades of coin-operated phonographs and Kinetoscopes. Library of Congress.

the average camera shot in American television ads lasted 3.8 seconds and the typical 30-second ad had an average of 7.9 shots. By the early 1990s, shot duration had dropped to 2.3 seconds while camera shots per ad had risen to 13.2. The goal was to deliver the message before the channel could be changed.[34] TV remotes, like mouse clicking on the Internet, have encouraged expectations of instant gratification. Moore's Law (the claim that the capacity of computer chips will double every two years) has become a kind of baseline expectation of improvement, and "slow" in the electronic realm has become an unalloyed measure of frustration and inferiority.[35]

## New Theme Parks and Thrill Machines

This double whammy of intensification and optimization in the on-going packaged pleasures revolution equally transformed the amusement park, that site of composite sensuality. Many of the rides in such facilities got faster and/or loopier, especially after the invention of under-track wheels in 1910 made it possible for coaster cars to negotiate sharp dips and turns without jumping off the track, creating the thrilling illusion of death-defying danger.

Of course there were limits to how fast, high, or twisty such rides might be without causing unbearable fear or nausea in the rider. In fact, coaster construction slowed with the Great Depression and World War II, and amusement parks fell into decline and disrepair, with only about two hundred remaining in the United States by 1960 — down from roughly ten times this number during the peak years of the 1920s.[36] One exceptional success story in this period, however, was Disneyland — erected in 1955 — notable for its founder's rejection of mere thrill rides. Walt Disney insisted instead on a seemingly new concept: a totalizing park environment with spectacles and rides designed to envelop the visitor in cinematic scenery and narratives. Disneyland provided a moviegoing public with a kind of fantasy immersion experience, placing the visitor at the scene of what, up to then, had been an experience found only on the silver screen. Disney also offered fantasy recreations of his own cultural and historical obsessions: his boyhood memory of small-town America ("Main Street"), storybook-like settings of African adventure, romanticized reconstructions of the American frontier, and futuristic fantasies of space travel and domestic convenience. Disney's own company in 1971 expanded and transported his scheme to central Florida with a whole new complex of parks, resorts, and shopping arcades called Walt Disney World.

Disney abandoned the logic of the "iron ride" in old amusement parks, and so instead of ratcheting up the thrill of motion, Disney took the tamer rides of the past (indoor scenic rail and simple carnival rides, for example) and "themed" them. Disney used the old technology of the mechanical swing, the rotating carousel, and the controlled roll of the coaster, but wove them into narratives — like the Dumbo the Flying Elephant ride — offering emotional associations with history, culture, and nostalgic memories of Disney cartoon characters. The attractions themselves were not as revolutionary as he and others claimed; indeed, they borrowed heavily from the themed and narrative-based attractions of early-twentieth-century parks — like Coney Island's Trip to the Moon or The Johnstown Flood. But Disney's attractions were adapted to suit contemporary sensibilities — the freak show and disaster themes of early Coney Island, for example, were eliminated. Disney also introduced new technologies to heighten the emotional impact: in the early years audio-animatronics, a magnetic tape-driven system to coordinate music with movements of mechanical characters, and more recently 3D movies. Disney thus induced customers to pay hundreds and eventually thousands of dollars to wait in long lines to experience Peter Pan's Flight, Mission to Mars, Honey, I Shrunk the Audience, and the Twilight Zone Tower of Terror. Disney's was a model of the calibrated, sensually complex "thrill" ride and site.[37]

Yet just a few years after Disneyland opened, others continued on the path of intensification. We see this in the striking revival of the amusement park and revival of the roller coaster and other "iron rides." This began in 1961 with Six Flags Over Texas (near Dallas), followed by Six Flag parks in Atlanta (1968) and St. Louis (1971). Imitations sprouted, including Magic Mountain in Valencia, California (1971), King's Island in Cincinnati, Ohio (1972), and Great America outside Chicago (1974), just as older parks underwent expansion — like Cedar Point in Ohio and Kennywood and Hersheypark in Pennsylvania. These parks did not simply return to the old "iron rides" of the past; instead, they often adopted Disney's approach of setting rides in idealized historical scenes. Six historical periods were represented by the different "flags" of the Six Flags attractions, and a simulated Yankee fishing village, Yukon mining camp, and New Orleans French Quarter were featured at Great America.[38] What really gathered the crowds, though, were the ever more thrilling coasters and water flumes. Whereas the old coasters had had tracks attached to wooden frames and elaborate scaffolding, a new generation of coasters from the 1960s traveled on tubular steel suspended on giant steel poles. Disneyland introduced the first of these in 1959 for its heavily themed Matterhorn, a ride featuring make-believe bobsleds running through a replica of the Swiss mountain. Quickly others followed, typically stripped of the expense and thematic aura of Disneyland, while also taking the potential for sheer motion thrills to new heights.[39]

Yet another level of thrill potential was realized with the further development in 1975 of a system of three sets of wheels fixed on and surrounding tubular steel track, allowing coasters to go upside down in a corkscrew route, for example. The ride could be faster and steeper (but also smoother), creating new sensations. "Positive gravity" resulted from a rapid acceleration up a high incline, making riders feel heavier. By contrast, the "negative g" experienced with a rapid downhill acceleration created the distinct sense of floating or even weightlessness — sensations resonating also with the much publicized reports of astronauts (both in space and in training). Within twenty years, there would be fifty varieties of patented, branded, roller coaster loops and turns on the market.[40] Coasters were advertised as imitating the experience of daredevil airplane pilots, with the resultant sensation of euphoria caused by the brain releasing dopamine. Chemicals of this sort were released as the brain was fooled into thinking the body to be at risk — and so sent distress signals to the heart, lungs, skin, and other vital organs, making the rider feel more aware, more "alive." The experience has been compared to the sense of well-being felt by athletes experiencing "runner's high."[41]

The new thrill rides quickly won devotees, especially from older kids and youth who thought they were too "cool" for Disney. Magic Mountain, which began as a rather sedate park built by Sea World developer George Millay in 1971, took off in 1976 with the introduction of the vertical-looping Revolution Free Fall (with a 55-mph drop in 2 seconds) and, later, the Colossus, noted for its height. By 2003, Magic Mountain was offering 16 different roller coasters, including Superman: The Escape, a ride utilizing a linear induction motor to create an acceleration of up to 4.5 times the force of gravity, reaching 100 miles per hour in 7 seconds. But older parks like Ohio's Cedar Point also joined the big-time coaster club, accumulating 16 thrill machines by 2003, including the astonishing Top Thrill Dragster, a "ride" that hydraulically launches riders up (and down) a 42-story tower. The Dragster reaches a speed of 120 miles per hour in less than 4 seconds, following which the rider experiences a brief bout of zero gravity before plunging back to earth.[42]

Even Disney has tried to reach the excitement-seeking crowd. In the 1990s, all Disney parks disguised thrill rides as "theme" experiences. Such is the Twilight Zone Tower of Terror at what is now called Disney Hollywood Studios in Orlando. Although really little more than a vertical-shaft ride, it is clothed as a runaway elevator set in a spooky, decrepit hotel. In 2006, even the innovative environmental zoo of Disney's Animal Kingdom introduced a themed coaster, Expedition Everest, with a none-too-persuasive backstory about riders pursuing a yeti.[43]

So it goes. In order to compete, sell, and profit, pleasure engineers packed more and more energy and taste in food and drink; greater volume, tone, and range in sound; more pictures and special effects and faster cuts in film and video games; and, of course, speedier and more body-defying sensations in thrill rides. In many cases this intensification purified the sensation, shearing away sensual complexity and sometimes social interaction. Except for Disney's traditionalism, the new thrill rides make little pretense to do anything more than offer "Gs."

Sometimes, however, intensity is not enough nor even desired by consumers. Merchandisers of packaged pleasures have learned to calibrate sensuality but, in so doing, have sometimes found a positive alternative to intensification. New ways have been developed to insinuate their packages into our daily routines, providing us with a push or pull of comfort or stimulation to smooth out the moods of the day and filling our worlds with constant sound and images. The packaged pleasure compounds sensuality by offering multiplicities of taste and flavor, not just jolts. It takes the form of richly complex environments of sight, smell, sounds, and motion that make people come back to Disney parks to recapture sensual surroundings and pleasures evoking memory and the promise of always something new.

The pleasures engineered and marketed in the last century of consumer culture have come in many packages containing diverse delights: soda, candy bars, cigarettes, phonographs, movies, and amusement park rides, just to mention those examples we have chosen to highlight. These innovations emerged largely between 1880 and 1910, but have taken off and expanded ever since. They all share in being based on the intensification and commercialization of sensations—taste, sight, sound, and motion—"plucked" from nature and from culture, and then packed, often labeled, delivered, and usually sold in personal portions. Pleasure engineers have of course reduced many of life's inconveniences and filled several of its remaining dull corners with sensation. But we still need to address in our final chapter how this revolution challenges our understanding of the goods that surround us, and what this revolution may mean for our future.

# Red Raspberries
# All the Time?

We live in a world of massive overconsumption, with the poorer countries of the globe racing to emulate the rich. The oversated have set this standard for the rest, that to enjoy the world is to capture it through consumption. Packaged pleasures make this new world possible, bringing new ways of containing, mobilizing, and selling the traditional sounds, sights, and oral satisfactions of human experience — and even our sense of motion and psychotropic imagination. This is the packaged pleasure revolution, a transformation in human sensuality so fundamental to human experience that it is easy to overlook — for it is right under our collective noses.

The revolution is partly — but only partly — about increasing wealth. New manufacturing technologies have allowed real prices to drop, making the rare into the commonplace, allowing former luxuries to become modern necessities. The revolution is therefore also about the democratization of consumption: the poor can now have goods once available only to the rich. And the revolution is also about new modes of distribution, and therefore vastly increased access. Goods now move far more freely about the planet, transcending seasonality and the limits of space. Networks of distribution allow commodities to penetrate the uttermost ends of the earth, bringing the goods (or bads) of city life to the countryside and vice versa, and nowadays even virtually, with the transport of images, sound and text of virtually every sort by electronic means: "telepresence." Precision marketing and engineering make these packaged pleasures far more desirable, recrafting while at the same time enlarging the appearance of "choice," a word that nicely conflates quality and our capacity to choose.

Another part of this revolution is the transformation of human sensation through novelties, inventions. The phonograph captures and eternalizes sound, just as the photograph captures and eternalizes sight—both of which are eventually "cheapened" as a result (for better or for worse). Other parts of this revolution are about new kinds of sensations or combinations (and transmogrifications) of traditional sensations. The amusement park combines sights and vertigo in novel forms while simultaneously miniaturizing and virtualizing travel; the candy bar, with novel combinations of textures, colors, and smells, fires neurons in different parts of the brain, accentuating its appeal beyond that of the carrot with its subtle taste and lack of significant smell. Advertisers add the compelling attraction of the wrapper and new fantasy accents, turning beer, soda, or cigarettes into expressions of identity attached to health, romance or adventure.

So what is the problem?

The problem is not so much the increased satisfaction of desire, per se, which at one level is nothing but the timeless human pursuit of "the good life." The need is not just for moderation, which overpersonalizes the predicament while underestimating the difficulties of self-control. The obesity epidemic is only partly about gluttony and sloth; it is also about the technological and commercial causes that make gluttony and sloth such easy choices. The problem is not just about affordability, since that is a basic function of economic success. Nor is it simply about access, since that is the virtue of democracy. All people, ideally, should have equal access to the fruits of civilization.

There is, however, a problem in that packaged products have been engineered, manipulated, and marketed for the singular purpose of maximizing sales and profits with so little regard for social and bodily consequences.[1] The problem is that the increased access to pleasurable goods (in commodity form) has tended to displace and subordinate pleasures that are not consumer goods. Keep in mind that the winning strategy for manufacturers has not necessarily been to go to extremes; the effort has rather been to integrate commodified sensory novelties into the push and pull of daily life, with the point always being to maximize profits.

Profits are not hard, of course, when the product is addictive. Financial guru and mega-investor Warren Buffett explained his purchase of a major cigarette manufacturer along these lines: "It costs a penny to make. Sell it for a dollar. It's addictive." And indeed, some packaged pleasures are, by virtue of how they are used and how they railroad the body, profoundly addictive. Nicotine is a physically addicting molecule, for example, albeit one whose addicting power is maximized when delivered in the inhaled form of the cigarette. Cigarette engineers have spent tens of billions of dollars to design a perfect engine of addiction, with advertising and packaging augmenting their attractiveness and impact. We tend to focus on the physiological properties of the addicting alkaloid and that is surely important, but we need also to appreciate how the means by which that molecule is delivered can be crucial to its appeal and use—and addictive potency. The same is true for opium or coca leaf, which become far more addictive when purified and injected or inhaled.

Clever packaging also prompts trial (including by kids) and discourages quitting or even moderation. This is the cigarette industry's notorious "bait and hook" strategy: the bait is the advertised image (including the attractive package) attached to a promise of looking glamorous or adult, followed by the hook, the rewiring of the brain by nicotine, making it difficult for a person to feel "normal" without their regular "hit." That is one reason cigarette companies are so adamant in resisting regulation of nicotine levels in cigarettes and laws restricting (or eliminating) package art. Without nicotine, the industry cannot create and sustain addiction. And without package art, the industry cannot create its fantasy world of naughty transgression so crucial to the appeal of kids first starting to smoke.

Cigarette manufacturers for years refused to admit that cigarettes were physiologically addictive, making a fetish of the "choice" made by smokers each time they light up, ignoring the crucial fact that consumer choice is often constrained by how such products have been designed. Product design is the "choice" we too-often ignore—and the blind spot in doctrines of consumer sovereignty. *Manufacturers* choose to make food or tobacco or amusement parks in certain ways, and consumers may find it hard or even impossible to deviate from that path. Captains of industry don't like us to think about this part of "choice"—choices made by manufacturers or regulatory authorities governing production. The focus is always on the consumer as a free agent, and the manufactured good as something born on the retail shelf (or "gift of God" in the words of makers of leaded gasoline).[2]

Other packaged pleasures are clearly habit forming, or even addictive to a variable extent. This includes junk foods with their multisensory appeal but also the highly engineered "hits" of sound, sight, and motion we have chronicled in this book. Modern people with or without the pains of withdrawal find it difficult to give up those engineered concoctions to develop more sociable pastimes or even to work and educate themselves when they can play video games or plug into iPods.[3]

Some such attractions can be explained by the lesson of Pavlov's dog: habits can be sparked by indirect stimuli. Behavioral psychologists have long known these stimuli to be cues associated with some pleasure source (the smell of doughnuts or a recorded song linked with emotion-evoking memories, for example). These cues, especially when multiple, create the intense desire to pursue the rewarding stimulus over and over. The consumer of packaged pleasures is cued to desire the "reward" inside—and repeated use results in an ingrained habit. A chain of cues then leads to cravings and further consumption, a vicious circle. This series of actions and reactions can become encoded in the brain in such a way as to lead to urges without much thought. And once an ingrained habit, the quest to satisfy some desire is no longer even required. "Conditioned hypereating" is the name David Kessler has given this process, but something similar can occur with nonfood packaged pleasures.[4]

In a nutshell, modern pleasure engineers have facilitated access to intensified sensations while also creating the "cues" that drive consumers to pursue these pleasures again and again. As a result, other kinds of desires are crowded out. Kessler notes that desire for "comfort food" (a linguistic abomination dating only from the 1980s) acts as a substitute for other emotions—"since it occupies working memory, and the brain can only focus on a limited amount of stimuli at any given time." Some people of course will be able to resist this pressure more easily than others, but the modern world's dazzling plenum of seductive packaging and advertising creates a "conducive environment" for succumbing to food, video games, or so many other packaged pleasures.[5]

By the very fact that engineers and marketers have been relentless in making these packaged pleasures more accessible, appealing, and ultimately habitual, much of the world beyond these packages has often become distant or even undesirable. Nature's attractions, after all, are often subtle and diffuse—lots of browns and greens and grays, with here and there a few red raspberries. But the package promises us raspberries all the time. And it can easily displace other, more subtle delights and experiences: candy bars drive out carrots, just as the hi-definition screen drives out the low, and convenience trumps toil. Thrill sights and rides displace the sublimity of strolls through, say, the English Lake District or quiet, even contemplative gazes across the Grand Canyon, especially insofar as such "attractions" are distant or otherwise difficult to access. And nature, like broccoli or amateur music, doesn't advertise. The continual mix of a thousand tunes on the iPod that "backsounds" the brains of millions as we work, travel, and play inevitably affects how we listen, or what we see or hear when not attached to some digital device. The packaged pleasure is often the path of least resistance.

Yet all this cannot be bad. Not many of us would give up the convenience of the most highly engineered forms of the package—smart phones, the extraordinary access to "information" (or whatever) via the Internet, and the flood of gadgets yet to come; neither will we abandon more prosaic packages—microwavable frozen food, canned beverages, or perhaps even that occasional Snickers bar. So where and how to draw the line? Cigarettes and other highly addictive killers are probably best put in a category by themselves, but many packaged pleasures have more mixed moral consequences. And efforts to challenge the bad and to sort these out from the acceptable have had a checkered history.

## A Bit More History: Past Efforts to Cork the Bottles

History is full of efforts to draw lines between acceptable and unacceptable pleasures. Early-nineteenth-century temperance advocates sought abstention from distilled drink, for example, but not from beer or wine. In fact, by 1850 there was a major shift in drinking fashions as imbibers abandoned spirits and wine for lager, a light amber beer that was relatively easy to ship, store, and preserve, especially when rice was added to the malted barley. This compromise did not last long. Saloons, often controlled by large brewers (Pabst, Schlitz, and Anheuser Busch, for example) turned into gathering spots for working class men and even criminals. In response, groups like the Women's Christian Temperance Union (founded in 1874) demanded the prohibition of all inebriating beverages as threats to family life—even as brewers attempted to rebrand beer as a healthful beverage for the family table.[6] Sometimes new psychotropic packages produced opposition that gave way with successful marketing. Calls to prohibit the sale of cigarettes were common in the early years of the twentieth century, and fifteen U.S. states actually succeeded, with the last of these bans (in Kansas) disappearing only in 1927. These challenges subsided in the wake of advertising campaigns designed to make smoking acceptable as a harmless pleasure. The temptation of tax revenue seems also to have been difficult to resist, and by the 1930s most states were profiting from cigarette taxes.[7]

Governance of other psychoactive substances has been very different. The 1906 Pure Food and Drug Act and the 1914 Harrison Narcotics Tax Act discredited and then regulated drugs such as cocaine and morphine and eventually even marijuana.[8] And of course Americans enacted a national ban on the sale of alcohol in 1919, with exceptions allowed for cider and homemade and sacramental wines. But cigarettes and alcohol (after the repeal of Prohibition in 1933) not only survived but flourished. Some of the illegal drugs had well-known addictive properties, but politicians and middle-class moralists also associated cocaine and marijuana with undesirable minorities and immigrants. Other potentially addictive drugs, including nicotine and caffeine, were left off the list. What was the difference? One key fact is that nicotine and caffeine were stimulants, and therefore seen as not really interfering with—but even helping—the ordinary citizen's ability to work. Both drugs also, by the 1910s, had powerful economic (and thus political) interests behind them. Coffee and tobacco were major consumer industries and widely viewed as helping to manage the moods of millions of Americans throughout the day. Defining the difference between socially acceptable desire and addiction was essential for caffeine, nicotine, and ultimately alcohol to flourish.[9]

Popular responses to sugared superfoods and overeating reveal similar patterns of ambivalence. As recently as 1900, the stout physique was treated as a sign of success, while thinness was a sign of sickness and poverty. But as high-calorie food and drink became cheaper and more plentiful, visible signs of being well-fed came to indicate not wealth but rather lack of self-control. The affluent were thus drawn to newly opened weight-loss "salons" and to countless faddish diets. As the popular magazine *Living Age* remarked in 1914, "Fat is now regarded as an indiscretion, and almost as a crime." Weight control became symbolic of a wider effort to rein in unconstrained desire. The battle against fat also became a way to recover or sustain youth, especially once it was realized that obesity causes premature aging. Fashion also made fat ever more obvious with the disappearance of corsets and the new vogue for more revealing clothing. Thinness became attractive, and makers of cigarettes (such as Lucky Strikes) capitalized on this, pushing smokes as a way for women to keep slim and trim: "Reach for a Lucky instead of a sweet!"[10]

Wilbur Atwater's techniques to quantify the calorie content of foods in 1894 provided a way of gauging the impact of any food on weight and led to a century of calorie counting. Dietary fads, including fasting, Fletcherism (the ritualistic hyperchewing of food), and even surgical fat removal, won and then lost support. But the goal of weight reduction was constant wherever the packaged pleasure had made inroads. Physicians began to take notice of degenerative diseases (rather than just the infectious ones) and the role that food played in making people sick. Thanks in part to the campaigns of John Harvey Kellogg against the evils of meat (as well as spices, tobacco, coffee, and cola), cereal began to replace meat at breakfast, while home economists promoted consumption of fresh fruit and vegetables. One consequence was a drop in the per-capita consumption of beef in the United States — from seventy-two pounds in 1899 to a mere fifty-five pounds in 1930. The identification of thinness with virtue drew on old Christian (and especially Puritan) values of restraint in the face of temptation, but the threat of temptation had a new urgency after 1900 with a growing awareness of the dangers of packaged pseudofoods. All of this might have led to a nation of slender people, and indeed, for the first three-quarters of the twentieth century, obesity was not a problem for most Americans.

Recent decades, however, have seen an explosive rise in morbid obesity, even as the threat from cigarettes has started to ease. Key factors in this rise are reduced exercise (from automobility and screen-induced sedentarism) but also a greatly increased consumption of snack and fast foods, especially foods containing high-fructose corn syrup.[11] High-fructose corn syrup is now added to literally hundreds of consumer products, thanks to tax advantages favoring it over sugar, and negligent oversight by the FDA (which still labels it "generally regarded as safe"). High-fructose corn syrup is a particularly dangerous pseudofood, and movements to limit its use are likely to grow.

Other packaged pleasures have resulted in similar efforts to rein them in. Self-proclaimed standard bearers of refinement attacked early movies, recorded music, and amusement parks, fearing cultural degeneration and overstimulation. For intellectual elites like Maxim Gorky, the concentrated sensuality of the amusement park tapped into the unrefined and even self-destructive longings of the laboring classes.[12] This was a common theme of early-twentieth-century disgruntlement, that the increased tempo and tedium of city life was leading laborers to seek overstimulation in their brief hours of leisure. Bruce Bliven in the *New Republic* wrote about visitors to Coney Island in 1921: "A mind buffeted by the whirlwind of life in New York, assaulted by the roar of machinery, excited by the pace at which we spin along, learns to regard a shout as the normal tone, and cannot hear with comfort anything less strident."[13]

As with hostility to drugs and gluttony, attacks on the amusement park often reflected the privileges and prejudices of elites (from old money or inherited status). Conservative elites worried that a rising commercial culture was subverting their old role of taste making. The genteel faced a losing battle with the Barnums, whose circuses, side-shows, movies, and popular recordings competed with libraries, play-grounds, and public education—all of which became the objects of re-formist philanthropy. Inevitably, these elites denounced the "popular amusements" and tried to regulate access to them.[14]

The genteel themselves avoided tacky Coney Island and had with-drawn from freak-show audiences by the early 1900s. But this did not mean they rejected the amusement park altogether. Middle-class re-treats included the more sedate seaside settings like Rye Beach and the family-friendly Playland amusement park on Long Island. And in the 1920s, impresarios of sited pleasure began to make such parks accept-able to middle-class families with rides and other attractions designed for small kids and cross-generational crowds (culminating in Disney-land in 1955). Reformers found various ways to censor films, especially when movies were exhibited in down-market nickelodeons. But again, these critics did not reject movies altogether. Instead, many bourgeois consumers joined the crowd at the new upscale movie palaces that had emerged by the mid-1910s. Soon thereafter many even embraced jazz recordings, especially in the whitened form of Paul Whiteman and his orchestra. These shifting boundaries were often governed by class prejudices that softened with time, but survive in surprising ways. By creating some limits to sensual indulgence, modern Americans gave themselves permission to enjoy other pleasures—alcohol and tobacco (off and on), junk foods, and Disneyland.[15]

Such confusion about restraining pleasure perturbed intellectuals when they attempted to grasp and confront the onslaught of the pack-age. The French sociologist Émile Durkheim feared that "the masses" would be enervated, even immobilized, by the assault on the senses so tantalizingly exhibited on billboards and in store windows.[16] Similarly, the Spaniard José Ortega y Gasset in his *Revolt of the Masses* (1930) argued that new shopping and amusement sections of cities were attracting uncouth crowds of free-spending but unsophisticated consumers. In his grim assessment, the "masses" with higher wages and more free time were imposing their tastes on the cultured minority, driving out refinement. This cultural version of Gresham's law would become central to generations of cultural jeremiads against "mass culture."[17]

Aldous Huxley's 1923 essay "Pleasure" and his classic *Brave New World* (1931) capped off this traditionalist perspective—portraying a world where people were not so much citizens as somnambulant consumers. From the feel-good drug soma ("all the advantages of Christianity and alcohol; none of their defects") and casual state-sanctioned sex to the "feelies" transcending "talkies" at the movies (he was repulsed by *The Jazz Singer*)—Huxley's dystopia would make people "love their servitude" to a social order that allowed neither personal thought nor initiative. Happiness in this brave new world meant reduction of tension, through satisfaction of all human senses at once, if possible. Freedom of thought was allowed only for the trustworthy elite. Readers of *Brave New World* today may find something of this cynical manipulation in our own era's consumer engineers, even without much state involvement.[18]

Fears such as these were hardly confined to defenders of a dying European culture. Educators encouraged the preservation of traditional games, folk songs and dances, and crafts in schools in hopes of counterbalancing the influence of novelty in commercial amusements.[19] John Dewey and his followers in America, for example, wanted schools to provide "education for life" as a counterweight to the allures of commercial pleasures, especially with the expectation of increased leisure time.[20] Critics of Coney Island also promoted local neighborhood playgrounds and public pools for children, along with public golf courses and tennis courts for adults.[21] But public recreation professionals (along with educators and librarians) knew what they were up against. Roosevelt's Research Committee on Social Trends in 1933 summed up the problem: "How can the appeals made by churches, libraries, concerts, museums and adult education for a goodly share in our growing leisure be made to compete effectively with the appeals of commercialized recreation?"[22]

These jeremiads surely missed the mark in many ways; Durkheim was wrong to fear that packaged pleasures would undermine social efficiency, and Huxley was wrong to think that totalitarian states would foster hedonism. They failed to fully realize how the modern engineering of desire has impacted everyone, not just the "masses." The problem runs deeper.

## What Is to Be Undone?

It should be clear by now that we do not place all packaged pleasures in the same box; we realize that different packages have very different impacts—and so merit different attitudes and approaches. Some, like cigarettes, are so addictive and deadly (and unloved) that they should probably not even be sold; indeed, courts have already ruled that cigarettes are defective and their manufacturers racketeers. Most other packaged pleasures pose lesser threats or practically none at all. Many give more and more frequent opportunities for fun to more and more people, releasing us from the daily grind. And some may even liberate users from the humiliation of being left out. Still, while some packaged pleasures have few downsides, many pose challenges that deserve a rethink.

So what about products based on high-fructose corn syrup or just plain sugar? A controversial proposal is the so-called Twinkie tax on sugars added to food and drink, first advocated in the mid-1980s by Kelly Brownell, a Yale psychologist, to discourage junk food consumption. Such a tax could also pay for programs to encourage healthy diets against the onslaught of powerful corporations selling processed, sugar-laden foods and drinks.[23] Support is growing for taxes of this sort, along with efforts to limit the container size or availability of soft drinks served in restaurants or schools. Some restaurant chains are even publishing calorie counts of various choices, though diners often find these hard to digest.

For other packaged pleasures, however, the moral imperatives are far less clear. Compression of sensuality, after all, heightens our range of experiences, offering us an amazing array of sights, sounds, and sensibilities that nature rarely (if ever) provides. Yet the sensual intensity of "the cinema of attractions" of the early twentieth century and the far more visually striking intensity of today's video games can make it easy to ignore or even disdain more diffuse and cultivated pleasures. A hike in the woods or an urban park does not overwhelm or exhaust us with an unremitting assault on the senses; instead, it delights us with the more subtle shades of complex life and, with experience, can provide us with a deepened recognition and understanding of undigested nature—along with fresh air and exercise. By contrast, the immersive and programmed character of pleasure packs (notably video games) can create psychological dependency. Social psychologists Jim Blascovich and Jeremy Bailenson point to research showing that some of the same reward centers of the brain are sparked by both drug use and less "physical" dependencies (like video games).[24]

We've explored how the accessibility of packaged pleasures offers us the convenience of sensual satiety in the form of treats once available (if at all) only to the rich. So ordinary people were overjoyed to be able to play Caruso records whenever they wished in 1910, just as they are eager to download and hear the latest techno song on mp3 players in the new millennium. In many ways the promise of access is what sold consumers on packaged pleasures in the first place, and keeps on selling them today. The Internet offers a virtually unlimited bacchanalia of sensory indulgences—for free or for purchase—with little more than a couple of mouse clicks or finger swipes.[25]

But convenience can also prejudice our priorities, increasing our likelihood of choosing the easiest, fastest entertainment or nourishment with the expectation of immediate gratification. Our point is not so much that "good things come to those who wait"; indeed, this rather fatalistic adage really only made sense in an age when people had to wait for crops to grow (to eat) or some ship to come in (to profit) or some holiday to arrive (to party). Modern technologies have erased this need to wait. Even so, instantaneous satisfaction often leads to rapid satiation or even boredom and, thereby, the need for new and even more frequent gratifications. Music critic Simon Reynolds points out that "today's boredom is not hungry, a response to deprivation; it is a loss of cultural appetite, in response to the surfeit of claims on your attention and time." The pleasure engineer's focus on immediate delivery can actually frustrate what psychologists and neurologists say they've found—that happiness comes not from frequent "hits" of pleasure but from extended engagements with anticipating, planning, and "winning" satisfactions—more journey and fewer destinations.[26]

Packaged pleasures risk also the privatization of what was once a social or even shared experience. Think about the individually packaged candy bar or can of soda versus the family meal—or even the different sociability of the Native American peace pipe versus the modern American-blend cigarette. Certainly there is some good in that trend toward having "a room of one's own," allowing an escape from the forced togetherness of the one-room cottage or confines of some obligate church or festival, with all the conformity entailed therein. Some packaged pleasures certainly assist in the realization of liberties that few of us would want to reverse. We may even favor the mp3 player precisely because it gives us the crowd experience without its distracting annoyances and inconveniences. But privatizing technologies can also lead to a loss of social opportunities, making us less skilled or comfortable in working with or even talking to others.[27] Could it be that the sounds from earbuds are so pure and enticing that we become less willing to leave the iPod for the unprogrammed world of conversation? No one is completely immersed in such pleasures—and many are used fruitfully to share and to network—but video games and mp3 players can create a solipsistic myopia that makes it harder to share values or physical contact.[28]

In a broader sense, packaged pleasures force questions about the consequences of their sheer abundance and ease of access. Because they have progressively become cheaper and are available in practically unlimited quantities, they quickly stack up. Choice and accumulation were keys to the original appeal of phonograph records, but their purchase created the problem of storing and accessing the collection. Abandoning the cumbersome cylinders for the easily storable disc was one early solution, as was the use of labeled album sleeves that could be readily retrieved from a stack or shelf. Far better was the mp3 player with its potential for seemingly endless accumulation of tunes and easy-to-find favorites or favored categories with the wonder of digital search and retrieval. But the wild escalation of the stockpile made possible by digitization has also led to a novel psychological problem—the now well-recognized dilemma of "overchoice" or "data smog," the phenomenon of seeming to drown in the sheer quantity of options. Internet sites like YouTube allow easy and immediate access to a near-infinite number of images in the form of video clips, vastly expanding "humanity's resources of memory" but also reducing the difference between past and present—and with digital "surfing" and multitasking producing an "experiential thin-ness," creating what Richard Foreman calls a "pancake people . . . spread wide and thin as we connect with that vast network of information accessed by the mere touch of a button."[29] Joining this is the fact that there are now few natural, fewer economic, and scarcely any clear moral limits to acquisition. This leads us to another form of accumulation: the modern phenomenon of collecting—of songs, movies, and games (especially from Internet downloads) of course but also dolls, toys, cars, and countless other objects that so often come to possess collectors (and hoarders). New commodity-centered communities are one result, but so is certain isolation from the noncognoscenti.[30]

This amassing raises the obvious question about how much any person, class, or nation should possess, especially when the packaged pleasure emerges in a context of growing inequality, resource depletion, and seemingly unstoppable carbon pollution. Details of this latter calamity have yet to unfold, but certainly one of its causes is the making of unrestrained consumption into something democratic and dignified—which also of course helps drive the economic machine upon which jobs and incomes depend. Perhaps we are not surprised that conservatives and libertarians will use "consumer choice" to rationalize continued expansion of the pleasure economy. But for many liberals, too, jobs and universal access to the "good life" justify the unlimited making, selling, and consuming of whatever can be stuffed into the malls and big box stores. Serious challenges to this consumer culture have long been confined to sermonizers on both the left and right.

So how can we enlarge our response? A beginning is to understand the scope of what has actually happened.

The meeting and expansion of the desires of most people in 1900, when joys were few and hard to attain, was a triumph of science and enterprise. The resulting revolution in engineered desire contributed to an amazing achievement—a broadly democratic/consumer capitalism that "delivered the goods" in the twentieth century. More goods meant more jobs and more ways of joining in (and breaking off from) consumerism. But what often worked for twentieth-century consumers may not work for twenty-first-century consumers.[31] This seemingly limitless array of commercial goods makes Americans appear to be global gluttons, slaves to desire, mocked by friends and despised or even terrorized by adversaries. And, what will happen when access to all this stuff spreads to the wider world, the rest of the seven billion? A future is hard to see apart from worsening environmental and energy crises, with much of that being from resources used to make packaged pleasures. How long can we continue along this path?

If overconsumption poses a threat to the future of humanity, we have no choice but to reconsider its role in our personal and political lives. We do of course have some choice. David Kessler argues that the first step to avoid the allures of superpacked food is to recognize our vulnerability to the cues that induce habitual overeating. The food industry may well have "cracked the code of conditioned hypereating and learned exactly how to manipulate our eating behavior," but this doesn't mean we are powerless to respond intelligently. We need to find new ways to de-emotionalize food, and to recognize that "conditioned hypereating is a biological challenge, not a character flaw."[32] And much of what can be said for food applies equally to other engineered and marketed "hits" of pleasure.

This leads us to the question of how (or whether) we introduce these delights to our young. And if hedonic contacts become habits early in life, this raises the knotty problem: which pleasures are developmentally healthy or harmless fun, and which displace the acquisition of valuable skills and appreciations?[33] Most scholars agree that the main downside to children's video games is that they rob the child of time that could be used to read, to command a musical instrument, to explore nature, or to otherwise play more creatively (and with bodily vigor).[34] Video games pose yet another problem, though, since unlike traditional toys and games they are not necessarily abandoned as children become adults. Video games often lack the markers of immaturity that tell the emerging adult it is time to put them down—and game designers and marketers have learned to capitalize on this. Many video games are also not like traditional card or board games that can be played only when a crowd can be assembled. Video games can usually be played by a solitary individual, whenever and wherever. No wonder some children have trouble finding time for school, work, or friends when they immerse themselves in a virtual world.[35]

Packaged pleasures affect the young in still other ways, with obesity being among the most obvious. This is generally regarded as a responsibility for parents and teachers, but it should be the concern of manufacturers and policymakers and broader society as a whole. An obvious point, perhaps, is that packaged pleasures should not be used as child quieters or babysitters; we need to give up the idea too often heard that video games are educational or suitable as rewards for good behavior (like candy or TV time).[36] And we should revisit the idea that briefly flourished in the 1960s and '70s that advertisers should be limited in their incessant drumbeat of selling snacks and thrills to kids.[37]

This leads us to the next question: how do we as adults cut back on the packaged pleasures to which we have become habituated? One possibility is to find new sources of satisfaction to displace the more debilitating effects. This can be as simple as substituting protein and fiber for sugar and fat (though every dieter knows this is easier said than done). New habits can work to short-circuit the sequence of cues that lead to the uncontrolled desire for candy, cigarettes, or video games. This may mean finding new rituals—especially social ones—that break the habits of insularity that so many packages encourage. A deliberate retraining of the senses can be effective; this can mean pursuing hobbies or adventures that increase in satisfaction with skill and experience. This suggests doing what packaged pleasures cannot—stretching and diffusing sensuality, making the anticipation and planning for some pleasure—and the preservation of its memory—as important as the "hit" itself. In this, old proponents of simplicity and refinement may have something to teach us. Retraining the body and senses is no easy task, especially if we don't want to become ascetics. And if this all sounds boring, that may be part of the "withdrawal" experience. (Remember that "boredom" itself was not even a word in common use until the packaged pleasures revolution.) In *The Happiness of Pursuit*, psychologist Shimon Edelman uses neuroscience to make an argument made by humanists for centuries: happiness is maximized in mental and physical activity, in the quest for experience, and in efforts to anticipate the future rather than in just realizing some goal. If all economy is an economy of time, as Adam Smith once insisted, then economy is part of the problem. Speed and convenience is what the packaged pleasure often brings, and that may be too much, too fast.[38]

Recall also that packaged pleasures have their deepest roots in the nature to which humans adapted prior to the rise of civilization. That biological world is where our ancestors needed to get pleasure and to get it fast, because it would not last. The engineers of sensuality have built on this and continue to do so. Natural pleasure is devoured, as are most packaged pleasures. Critics of the rush to satisfaction have a point — the technologies of modern pleasure mostly (but not entirely) reinforce the natural rather than the nurtured. It is not that we need to abandon the fast pleasure, but rather to clear some space, time, and desire for the slower delights that require cultivation. The savored meal, the sharing of song, the exploration of a landscape (painted or real), the engagement in sports or hobbies or convivial parties all challenge the logic of the packaged pleasure.

A complementary challenge is to pursue "situated pleasures . . . grounded in the historicity of sensory experience," as Wendy Parkins and Geoffrey Craig suggest in their book, *Slow Living*. By contrast, packaging is all about taking pleasure out of some prior social, natural, and historical context. Reversing this trend, of course, is the point of taking kids to farms and on nature vacations — to see where things come from and how they are connected, and to learn thereby how to appreciate the wonders of life without artifice. The language of the slow food and related movements may be romantic and their proscriptions sometimes opaque, but Carl Honoré's idea of a "calm, careful, receptive, still, intuitive, unhurried, patient, reflective" approach to life can be a corrective, if not a replacement, for fast living.[39]

It is sometimes observed that there has been no net increase in those Americans who claim that they are very happy (35 percent in 1980 vs. 34 percent in 2006, according to one survey). Americans have long embraced the Enlightenment's dictum that happiness is attainable on earth, indeed that pursuit of happiness is a fundamental right.[40] And that helps explain why Americans are more apt to claim to be happy than other people. Yet Americans and many other people don't seem to be getting any happier over time.[41] Advances in material comforts, including packaged pleasures, don't seem to have made much difference in this respect. Some packaged pleasures have clearly locked us into habits that many of us wish we didn't have, and some even seem to have blinded us to other experiences we might prefer. We need to recognize what the package has done for us, but also to us, and to look for pleasures beyond its confines.

**Notes**

## Chapter 1

1.  Diane Ackerman, *Natural History of the Senses* (New York: Vintage, 1990), 27, 6, 82–83; Peter Farb and George Armelagos, *Consuming Passions: The Anthropology of Eating* (Boston: Houghton Mifflin, 1980), 22; Paul Rozin, "Sweetness, Sensuality, Sin, Safety, and Socialization: Some Speculations," in *Sweetness*, ed. John Dobbing (London: Springer, 1987).

2.  Daniel E. Lieberman, "Evolution's Sweet Tooth," *New York Times*, June 5, 2012.

3.  Microstock Posts, http://www.microstockposts.com/ever-wondered-how -many-images-are-on-the-internet/. YouTube videos have gone from 6 million in 2006 to several hundred million in 2013, with six billion hours uploaded monthly.

4.  Constance Classen, *Worlds of Sense: Exploration the Senses in History and Across Cultures* (London: Routledge, 1993), 27–29; Farb and Armelagos, *Consuming Passions*, 40; David Howes, ed., *Sensual Relations: Engaging the Senses in Culture and Social Theory* (Ann Arbor: University of Michigan Press, 2003), xix, 6–7, 44.

5.  Lionel Tiger, *Pursuit of Pleasure* (Boston: Little, Brown, 1992), 21–64.

6.  James Owen, "Bone Flute Is Oldest Instrument, Study Says," *National Geographic News*, June 24, 2009, 1–2.

7. Ackerman, *Natural History*, 202–6; Deryck Cooke, *The Language of Music* (New York: Oxford University Press, 1987), 31; Steven Mithen, *The Singing Neanderthals: The Origins of Music, Language, Mind, and Body* (Cambridge, MA: Harvard University Press, 2007), 234; B. Bower, "Doubts Aired Over Neanderthal Bone 'Flute' (and reply by Musicologist Bob Fink)," *Science News* 153 (April 4, 1998): 215.

8. Ackerman, *Natural History*, 213–14; William McNeill, *Keeping Together in Time: Dance and Drill in Human History* (Cambridge, MA: Harvard University Press, 1995), 4.

9. Barbara Ehrenreich, *Dancing in the Streets: A History of Collective Joy* (New York: Metropolitan Books, 2006) 22–23, 184, 210–15.

10. Ackerman, *Natural History*, 218.

11. Ibid., 142–46; Philipa Pullar, *Consuming Passions: Being an Historic Inquiry into Certain English Appetites* (Boston: Little, Brown, 1970), 19, 87; Massimo Montanari, *Food is Culture* (New York: Columbia University Press, 2006), 11–63.

12. R. Benders, *A History of Scent* (London: Hamish Hamilton, 1972), 91; Pullar, *Consuming Passions*, 87.

13. Alan Hunt, *Governance of the Consuming Passions: A History of Sumptuary Law* (New York: St. Martin's Press, 1996), chap. 1; Farb and Armelagos, *Consuming Passions*, 151–52; Massimo Montanari, *Culture of Food* (Cambridge: Blackwell, 1994), 82–87.

14. Simon Schama, *Embarrassment of Riches: An Interpretation of Dutch Culture in the Golden Age* (New York: Knopf, 1987), 125.

15. Carolyn Korsmeyer, *Making Sense of Taste* (Ithaca, NY: Cornell University Press, 1999), 16–33; Montanari, *Culture of Food*, 29, 59; Anthony Synnott, "Puzzling over the Senses: From Plato to Marx," in Howes, *The Varieties of Sensory Experience* (Toronto: University of Toronto Press, 1991), 61–71.

16. Ehrenreich, *Dancing in the Streets*, 87, 122, 138, 149, 179; Emmanuel Le Roy Ladurie, *Carnival in Romans* (New York: George Braziller, 1979), 178–80, 313; Peter Burke, *Popular Culture in Early Modern Europe* (New York: Harper, 1978), chaps. 7 and 8.

17. Ackerman, *Natural History*, 148–51. David A. Kessler, *The End of Overeating: Taking Control of the Insatiable American Appetite* (New York: Rodale, 2009), develops this in a powerful popular book; compare also the YouTube video by Robert Lustig, "Sugar: The Bitter Truth," July 30, 2009, http://www.youtube.com/watch?v=dBnniua6-oM, with 4.3 million views as of February 2014.

18. James Brook and Iain Boal, *Resisting the Virtual Life: The Culture and Politics of Information* (San Francisco: City Lights Books, 1995).

19. Farb and Armelagos, *Consuming Passions*, 17, 84; Paul Bousfield, *Pleasure and Pain*, (1926; New York: Routledge, 1999), 20–95; Michael Dietler, "Food, Identity, and Colonialism," in *The Archaeology of Food and Identity*, ed. Katheryn Twiss (Carbondale: Southern Illinois University, 2007), 222; Jim Drobnick, "Olfactocentrism," in *The Smell Culture Reader*, ed. Jim Drobnick (Oxford: Berg, 2006), 1–7; Pasi Falk, *The Consuming Body* (London: Sage, 1994); Robert Rivlin and Karen Gravelle, *Deciphering the Senses: The Expanding World of Human Perception* (New York: Simon and Schuster, 1984), 9.

20. Émile Durkheim, *The Division of Labor in Society* (1893; New York: Free Press, 1964), 17, 353–73; Robert S. Baker, *Brave New World: History, Science, and Dystopia* (Boston: Twayne, 1990), 10; Aldous Huxley, *Brave New World* (1932; New York: Harper and Row, 1946), xvii–xviii, 284.

21. This is a vast literature and will be discussed in more detail in the conclusion.

22. The sociological roots of this populism are shown in Simon Patten, *The Theory of Prosperity* (New York: Macmillan, 1902), 230–31, with commentary by Daniel Horowitz, *The Morality of Spending: Attitudes Toward the Consumer Society in America, 1875–1940* (Baltimore, MD: The Johns Hopkins University Press, 1985), 3–44. Among the neoclassical proponents are George Stigler and Gary Becker, "De Gustibus Non Est Disputandum," *American Economic Review* 67 (1977): 76–90; and Stanley Lebergott, *Pursuing Happiness* (Princeton, NJ: Princeton University Press, 1993), chaps. 1 and 6.

## Chapter 2

1. Peter Farb and George Armelagos, *Consuming Passions: The Anthropology of Eating* (Boston: Houghton Mifflin, 1980), 4, 145, 161–69; Lionel Tiger, *Pursuit of Pleasure* (Boston: Little, Brown, 1992), 33.

2. A. Hausner, *The Manufacture of Preserved Foods and Sweetmeats* (London: Scott, Greenwood & Son, 1912), 111, 123, 149; Sue Shephard, *Pickled, Potted and Canned: The Story of Food Preserving* (London: Headline, 2000), 178–90.

3. Faber and Armelagos, *Consuming Passions*, 52; Alec Davis, *Package and Print: The Development of Container and Label Design* (London: Faber and Faber, 1967), 43; Eleanor Godfrey, *The Development of English Glassmaking, 1560–1640*

(Chapel Hill: University of North Carolina Press, 1975), 157-59; Edwin Morris, *Fragrance: The Story of Perfume from Cleopatra to Chanel* (New York: Scribner's Sons, 1984), 62-63, 136-39.

4. Agriculture spread eastward across the Bering Strait and west into Europe as humans depleted and destroyed the grass-eating beasts that had long stored energy in their flesh for human hunters. While farming may not have been the fault line formerly attributed to it (settlements may have predated the domestication of plants and animals), agriculture and the accumulation that it hastened certainly encouraged containerization. Clive Gamble, *Origins and Revolutions: Human Identity in Earliest Prehistory* (Cambridge: Cambridge University Press, 2007), 198-203, 275-76. Gamble also points out that pots created a kind of external stomach, just as blade and edge tools created a kind of artificial set of teeth and nails.

5. Brian Hayden, "The Emergence of Prestige Technologies and Pottery," in *The Emergence of Pottery: Technology and Innovation in Ancient Societies*, ed. William Barnett and John Hoopes (Washington DC: Smithsonian Institution Press, 1995).

6. Jean-Pierre Warnier, "Inside and Outside: Surfaces and Containers," in *Handbook of Material Culture*, ed. Christopher Tilley et al. (London: Sage, 2006), 192; Cameron B. Wesson, "Chiefly Power and Food Storage in Southeastern North America," *World Archaeology* 31 (1999): 145-64; Pierre Bourdieu, *Outline of a Theory of Practice* (Cambridge: Cambridge University Press, 1977), 195.

7. Marijke van der Veen, "Food as an Instrument of Social Change: Feasting in Iron Age and Early Roman Southern Britain," in Twiss, *The Archaeology of Food and Identity*, ed. Katheryn Twiss (Carbondale: Southern Illinois University Press, 2007), 112-29; Michael Dietler, "Feasts and Commensal Politics in the Political Economy: Food, Power, and Status in Prehistorical Europe," in *Food and the Status Quest: An Interdisciplinary Perspective*, ed. Polly Wiessner and Wulf Schiefenhövel (Oxford: Berghahn, 1996).

8. Michael Pollan, *The Omnivore's Dilemma* (New York: Penguin, 2006), 294-95.

9. Massimo Montanari, *The Culture of Food* (Cambridge: Blackwell, 1994), 5-11; William Longacre, "Why Did They Invent Pottery Anyway?" in Barnett and Hoopes, *The Emergence of Pottery*, 266-78.

10. Tom Standage, *A History of the World in 6 Glasses* (New York: Walker, 2005), 27-121; Michael Dietler, "Consumption, Agency, and Cultural Entanglement: Theoretical Implications of a Mediterranean Colonial Encounter," in

*Studies in Culture Contact: Interaction, Culture Change, and Archaeology*, ed. James G. Cusick (Carbondale: Southern Illinois University Press, 1998); Patrick E. McGovern, "Vin Extraordinaire: Archaeochemists Sniff Out The Oldest Wine in the World," *Sciences* 36 (1996): 27–31; Gregory McNamee, *Movable Feasts: The History, Science, and Lore of Food* (Westport, CT: Praeger, 2007), 91–92.

11. Fermentation also produced carbon dioxide for bread dough and lactic acid for cheese; it was also critical to making sauerkraut and cured meats. See Montanari, *The Culture of Food*, 11–17; Standage, *A History of the World in 6 Glasses*, 27.

12. Dietler, "Consumption, Agency, and Cultural Entanglement," 303; Shephard, *Pickled, Potted and Canned*, 125–26; James McWilliams, *A Revolution in Eating: How the Quest for Food Shaped America* (New York: Columbia University Press, 2005), 247–52.

13. Mandy Aftel, "Perfumed Obsession," in *The Smell Culture Reader*, ed. Jim Drobnick (Oxford: Berg, 2006).

14. Morris, *Fragrance*, 18–30; David Courtwright, *Forces of Habit: Drugs and the Making of the Modern World* (Cambridge, MA: Harvard University Press, 2001), 18, 72–74; Wolfgang Schivelbusch, *Tastes of Paradise* (New York: Vintage, 1992), 159.

15. Edward Emerson, *Beverages, Past and Present*, vol. 2 (New York: Putnam's Sons, 1980), 23–35; Michael Pollan, *Botany of Desire* (New York: Random House, 2001), chap. 1.

16. Morris, *Fragrance*, xiv, 3–5; Courtwright, *Forces of Habit*, 72–73, 243–45.

17. Hans Teuteberg, "The Birth of the Modern Consumer Age," in *Food, The History of Taste*, ed. Paul Freedman (Berkeley: University of California Press, 2007), 234–37.

18. Schivelbusch, *Tastes of Paradise*, 152–59; Courtwright, *Forces of Habit*, 75; Jad Adams, *Hideous Absinthe, A History of the Devil in a Bottle* (Madison: University of Wisconsin Press, 2004), 19, 58, 88, 124, 196–225.

19. Daniel Lord Smail, *On Deep History and the Brain* (Berkeley: University of California Press, 2008), 164–83; Roy Porter and Dorothy Porter, *In Sickness and in Health: The British Experience, 1650–1850* (London: Fourth Estate, 1988), 217–20.

20. Courtwright, *Forces of Habit*, 32–38.

21. Peter Pormann and Emilie Savage-Smith, *Medieval Islamic Medicine* (Washington DC: Georgetown University Press, 2007), 115–38.

22. Patricia Rosales, "A History of the Hypodermic Syringe, 1850–1920s" (PhD diss., Harvard University, 1997).

23. Rosales, "A History of the Hypodermic Syringe," 1–40, 98, 114; Kit Barry, *Advertising Trade Cards* (Brattleboro, VT: self-published, 1981), 13.

24. Henry T. Brown, *507 Mechanical Movements: Mechanisms and Devices* (1901; New York: Dover, 2005); Susan Strasser, *Satisfaction Guaranteed: The Making of the American Mass Market* (Washington DC: Smithsonian Institution Press, 1989), 7.

25. M. (Nicholas) Appert, *The Art of Preserving All Kinds of Animal and Vegetable Substances* (London: Black, Parry, 1811), 1–19; Clarence Francis, *A History of Food and its Preservation* (Princeton, NJ: Princeton University Press, 1937).

26. Francis, *A History of Food*, 9–11; Appert, *The Art of Preserving*, 1–19.

27. Arthur Judge, *A History of the Canning Industry* (Baltimore, MD: Canning Trade Association, 1914), 13–14; E. D. Fischer, *Everything for Canners* (Franklinville, NY, 1891), 1–3; T. N. Morris, "Management and Preservation of Food," in *A History of Technology*, vol. 5, ed. C. Singer (Oxford: Clarendon, 1958); Alfred Lief, *A Close-Up of Closures: History and Progress* (New York: Glass Container Manufacturers Institute, 1965); Kenneth R. Berger, *A Brief History of Packaging* (New York: Morris, 1958); chap. 3; Robert Opie, *The Art of the Label: Designs of the Times* (Secaucus, NJ: Chartwell, 1987).

28. Tom Hine, *The Total Package* (New York: Harper, 1995), 69–70; Waverley Root and Richard de Rochemont, *Eating in America, A History* (New York: Morrow, 1976), 158–59; Douglas Collins, *America's Favorite Food: The Story of Campbell Soup Company* (New York: Harry N. Abrams, 1994), 23–24.

29. Hine, *The Total Package*, 72; Hausner, *The Manufacture of Preserved Foods*, 4; James H. Collins, *The Story of Canned Foods* (New York: Dutton, 1924), 95–104; Richard Cummings, *The American and His Food* (1940; New York: Arno, 1970), 66–69.

30. William Cronon, *Nature's Metropolis: Chicago and the Great West* (New York: Norton, 1991); Richard White, *Railroaded: The Transcontinentals and the Making of Modern America* (New York: Norton, 2011).

31. Richard Cummings, *The American and His Food*, 61–64; Root and de Rochemont, *Eating in America*, 208.

32. Susan Williams, *Savory Suppers and Fashionable Feasts: Dining in Victorian America* (New York: Pantheon, 1985), 114; John Hoenig, "A Tomato for all Seasons," *Business History Review* (forthcoming).

33. Collins, *The Story of Canned Foods*, 18–48; Judge, *A History of the Canning Industry*, 54–56, 92–93; Cox Brothers and Co., *Canning Machinery* (Bridgeton,

NJ: Cox Brothers, 1900); Strasser, *Satisfaction Guaranteed*, 7, 32; Cummings, *The American and his Food*, 67-68; Fischer, *Everything for Canners*, 6-11; National Canner's Association, *Seventh Annual Convention, 1914* (Baltimore, 1914), 7.

34. Collins, *The Story of Canned Foods*, 24-48; Judge, *A History of the Canning Industry*, 83-93; *Bottle, Jar and Can Filling Machinery* (Chicago: Charles L. Bastian Co., 1904); Ernest Schwaab, *Secrets of Canning* (Baltimore, MD: John Murphy, 1899), 45-47. A good chronology of canning, written to aid in the archaeologic identification of old can dumps, is "How Old Is 'Old'? Recognizing Historical Sites and Artifacts," http://www.fire.ca.gov/resource_mgt/archaeology /downloads/Cans.pdf.

35. Walter Soroka, *Fundamentals of Packaging Technology*, 2nd ed. (New York: Institute of Packaging Professionals, 2000); Hine, *The Total Package*, 65-72; Davis, *Package and Print*, 83.

36. Lief, *A Close-Up of Closures*, 56-58. For details on mason jars, see http:// www.answers.com/topic/history-of-packaging-and-canning.

37. Thomas Chester, *Carbonated Beverages* (New York, 1882), 40-80; Beverage World, *Coke's First 100 Years* (Great Neck, NY: Beverage World, 1986), 75-124; John Newberg, *Crowns: The Complete Story*, 3rd ed. (Paterson, NJ: Lent & Overkamp, 1961), 1-2.

38. Beverage World, *Coke's First 100 Years*, 50; Newberg, *Crowns*, 99-101.

39. Beverage World, *Coke's First 100 Years*, 124-43; Richard Tedlow, *New and Improved: The Story of Mass Marketing in America* (Cambridge, MA: Harvard Business School Press, 1996), 112; Newberg, *Crowns*, 138; Hine, *The Total Package*, 74; Davis, *Package and Print*, 46-47; Crawford Johnson, *Coca Cola Bottling Company* (New York: Newcomen Society, 1987).

40. Strasser, *Satisfaction Guaranteed*, 31; Hine, *The Total Package*, 60; Berger, *A Brief History of Packaging*.

41. Hine, *The Total Package*, 61-62; Davis, *Package and Print*, 66; "A History of Packaging," *Ohio State University Fact Sheet*, http://ohioline.osu.edu/cd -fact/0133.html.

42. Scott Bruce and Bill Crawford, *Cerealizing America: The Unsweetened Story of American Breakfast Cereal* (Boston: Faber and Faber, 1995), 20-29, 50; Gerald Carson, *Corn Flake Crusade* (New York: Rinehart, 1957), 162-70; David Goodman and Michael Redclift, *Refashioning Nature: Food, Ecology and Culture* (London: Routledge, 1991), 34-35; Andrew Smith, *Eating History* (New York: Columbia University Press, 2009), 123-26, 141-52.

43. Stephen Fenichell, *Plastic: The Making of a Synthetic Century* (New York: HarperCollins, 1996), chap. 3; Berger, *A Brief History of Packaging*, 56-58.

44. Fenichell, *Plastic*, chap. 5; George Borgstrom, "Food Processing and Packaging," in *Technology in Western Civilization*, vol. 2, ed. Melvin Kranzberg and Caroll W. Pursell (New York: Oxford University Press, 1967); Hyla M. Clark, *The Tin Can Book: The Can as Collectible Art, Advertising Art, and High Art* (New York: New American Library, 1977); Morris, "Management and Preservation of Food"; Opie, *The Art of the Label*, chap.3; Stuart Thorne, *The History of Food Preservation* (Kirby Lonsdale: Parthenon, 1986).

45. Charles L. Van Noppen, *Death in Cellophane* (Greensboro, NC: by author, 1937).

46. Richard Franken and Carroll Larrabee, *Packages that Sell* (New York, Harper, 1928), 1-11.

47. Clayton Smith, *The History of Trade Marks* (New York: self-published, 1923), 9-11; Root and de Rochemont, *Eating in America*, 158-59; Opie, *The Art of the Label*, 1-9.

48. Strasser, *Satisfaction Guaranteed*, 19, 30.

49. Ibid., 44-45.

50. Robert Jay, *Trade Cards in Nineteenth Century America* (Columbia: University of Missouri Press, 1987), 20-44, 69-74; David Cheadle, *Victorian Trade Cards* (Paducah, KY: Collector Books, 1996), 13; Allan Brandt, *The Cigarette Century: The Rise, Fall, and Deadly Persistence of the Product That Defined America* (New York: Basic, 2007), 19-44.

51. Strasser, *Satisfaction Guaranteed*, 53, 91, 164-67.

52. Herbert Hess, *Productive Advertising* (Philadelphia: Lippincott, 1915), 25-35, 111-15; Herbert Hess, *Advertising: Its Economics, Philosophy and Technique* (Philadelphia: Lippincott, 1931), 154-95.

53. Charlene Elliott, "'Consuming the Other': Packaged Representations of Foreignness in President's Choice," in *Edible Ideologies: Representing Food and Meaning*, ed. Kathleen LeBesco and Peter Naccarato (Albany, NY: SUNY Press, 2008), 181-82; W. A. Dwiggins, *Layout in Advertising* (New York: Harper, 1928), 117.

54. Matthew Luckiesh, *Light and Color in Advertising and Merchandising* (New York: Van Nostrand, 1923), 1, 22, 39, 56-59; Franken and Larrabee, *Packages that Sell*, 45, 91-95, 157; J. Z. De Camp, "The Influence of Color on Apparent Weight," *Journal of Experimental Psychology* 2 (October 1917): 51; James Rice,

*Packaging, Packing and Shipping* (New York: American Manufacturer's Association, 1936), 119–25.

55. Opie, *The Art of the Label*, 52–53; Hine, *The Total Package*, 50; Tedlow, *New and Improved*, 15.

56. Historian Susan Strasser notes that the alcohol in Pinkham's patent medicine served as a solvent, extraction agent, and preservative, and that these medicines sometimes provided relief to customers. Belle Waring, "NLM Seminar Focuses on 19th-Century Patent Medicine" *NIH Record*, April 6, 2007, http://nihrecord .od.nih.gov/newsletters/2006/04_07_2006/story04.htm; Susan Strasser, "Commodifying Lydia Pinkham: The Woman, The Medicine, The Company," *American College of Obstetricians and Gynecologists Clinical Review* 12 (2007): 13–16. See also John Haller and Robin Haller, *Physicians and Sexuality in Victorian America* (Urbana: University of Illinois Press, 1974), 285–98; Cheadle, *Victorian Trade Cards*, 163; Jay, *Trade Cards in Nineteenth Century America*, 20–40, 44; Barry, *Advertising Trade Cards*, 33–34, 64–69; Jim Heimann, *All-American Ads, 1900–1919* (Los Angeles: Taschen, 2005), 51.

57. See, for example, Trade Cards and Labels Collection, Col. 9, box 3 and 8, Winterthur Library, Winterthur, Delaware.

58. Mark Pendergrast, *For God, Country and Coca-Cola* (New York: Charles Scribner's Sons, 1993), 199.

59. Pamela Laird, *Advertising Progress: American Business and the Rise of Consumer Marketing* (Baltimore, MD: Johns Hopkins University Press, 1998), 253–54; Hine, *The Total Package*, 78, 91–267; Bruce and Crawford, *Cerealizing America*, 67; Arthur Marquette, *Brands, Trademarks and Good Will* (New York: McGraw Hill, 1967), 46; M. M. Manning, *Slave in a Box: The Strange Career of Aunt Jemima* (Charlottesville, University of Virginia Press, 1998).

60. Collins, *America's Favorite Food*, 30–69, 89, 121–22.

61. Franken and Larrabee, *Packages that Sell*, chap. 3; Hine, *The Total Package*, 107; Tedlow, *New and Improved*, 15; Elliot, "Consuming the Other," 183; Carl Greer, *Buckeye Book of Direct Advertising* (Hamilton, OH: Beckett Paper Company, 1925), 5.

62. Tedlow, *New and Improved*, 27, 50–54; "Blazing the Trail," in *Report of Sales and Advertising Conference of the Bottlers of Coca-Cola 11*, Coca Cola Archives, March 27, 1923, cited in Tedlow, *New and Improved*, 34; Strasser, *Satisfaction Guaranteed*, 48–51; Beverage World, *Coke's First 100 Years*, 62, 176–81.

63. Strasser, *Satisfaction Guaranteed*, 35, 356–57; Harvey Levenstein, *Revo-*

*lution at the Table: The Transformation of the American Diet* (New York: Oxford University Press, 1988), 38–41.

64. T. J. Jackson Lears, *Fables of Abundance: A Cultural History of Advertising in America* (New York: Basic, 1994), chap. 4.

65. Uneeda ads: *Harper's Weekly*, July 31, 1909, 33, and September 9, 1911, 3; *Youth's Companion*, April 24, 1902, 217, and May 8, 1902, 239; *Ladies' Home Journal*, January 1903, 18; *Collier's*, October 8, 1910, back cover. Ad for the Westfield Book of Pure Foods (sponsored by and featuring Karo, Crisco, Kellogg, Knox Gelatine, Baker's Vanilla, and Beechnut foods), *Collier's*, September 27, 1913, 454.

66. John Lee, *How to Buy and Sell Canned Foods* (Baltimore, MD: The Canning Trade, 1926), 11–13, 16, 21.

67. Schlitz ads, *Harper's Weekly*, May 21, 1904, back cover, and June 28, 1903, 837.

68. Welch ads, *Collier's*, April 23, 1908, 33; May 14, 1910, 32; May 24, 1913, 11.

69. Ads for and articles about the Westfield Book of Pure Foods, *Collier's*, April 12, 1913, 38; June 7, 1913, 38; October 2, 1915, 13; "Penny Poisons," *Collier's*, October 25, 1915, 13; Lewis Allyn, "The Dealer Speaks," *McClure's*, November 1915, 38.

70. Van Camp ads: *Collier's*, June 26, 1909, 25; *Ladies' Home Journal*, May 1910, 55; Heinz ad, *Youth's Companion*, May 31, 1906, back cover.

71. Colgate ad, *Collier's*, May 22, 1909, 13; Gillette ad, *Collier's*, April 15, 1908, 35; Dr. Eliot's book ads, *Collier's*, April 3, 1915, 30, and August 14, 1915, 28.

72. Quaker oats puffed cereal ads: *Collier's*, May 27, 1911, 24; August 21, 1915, 26; *Ladies' Home Journal*, June 1909, 66; Van Camp ad, Feb. 9, 1911, 79; Campbell's soup ad, *Collier's*, September 4, 1915, 34.

73. Campbell's soup ad, *Ladies' Home Journal*, August 1916, 31; Borden's milk ad, *Collier's*, October 23, 1909, 33; Quaker oats puffed cereal ads, *Collier's*, February 21, 1914, 22, and November 8, 1913, 25.

74. Campbell's soup ads: *Collier's*, November 1, 1913, 21, and April 17, 1915, 19; *Ladies' Home Journal*, January 1910, 44, and June 1916, 41.

75. Hunt's California fruits ad, *Ladies' Home Journal*, January 1913, 1; Welch's grape juice ad, *Collier's*, January 17, 1914, 25; Snider's catsup ad, *Ladies' Home Journal*, January 1913, 65; Dole's pineapple juice ad, *Ladies' Home Journal*, January 1910, 69; Libby's fruit ad, *Youth's Companion*, September 17, 1914, 489.

76. Cream of Wheat ad, *Ladies' Home Journal*, October 1913, 64; Horlick's

malted milk ads: *Youth's Companion*, March 26, 1903, 156, and February 23, 1905, 97; *Collier's*, June 11, 1912, 4.

77. Trade Cards and Labels Collection, Col. 9, Box 7, Winterthur Museum Library; Miriam Formanek-Brunell, *Made to Play House: Dolls and the Commercialization of American Girlhood, 1830–1930* (New Haven, CT: Yale University Press, 1993), 90–92, 109–16; Joleen Robinson, *Advertising Dolls* (Paducah, KY: Collector Books, 1980) chaps. 1–2; Gary Cross, *The Cute and the Cool* (New York: Oxford University Press, 2004), chap. 3.

78. F. E. Ruhling, "How Cracker Jacks Keeps Itself in the Lime Light," *Candy Factory*, June 1921, 25–26.

79. Quaker oats ad, *Youth's Companion*, September 17, 1914, back cover.

80. Quaker oats ads, *Collier's*, August 21, 1915, 23, and June 5, 1915, 21; Post Toasties ad, *Ladies' Home Journal*, July 1909, 40; similar appeals to the desires of children appeared in ads for Kellogg's corn flakes; Kellogg's corn flakes ads: *Collier's*, July 11, 1908, back cover; *Ladies' Home Journal*, July 1910, 61, and March 1911, back cover; *Youth's Companion*, December 15, 1910, 707.

## Chapter 3

1. Claude E. Teague Jr. (R. J. Reynolds), "Research Planning Memorandum on the Nature of the Tobacco Business and the Crucial Role of Nicotine Therein," April 14, 1972, *Legacy Tobacco Documents Library*, University of California, San Francisco, http://legacy.library.ucsf.edu/tid/ryb77a00.

2. Jordan Goodman, *Tobacco in History: The Cultures of Dependence* (London: Routledge, 1993), 149–204. British consumption of tobacco in the eighteenth century was about two pounds per person per year; see Robert C. Nash, "The English and Scottish Tobacco Trades in the Seventeenth and Eighteenth Centuries," *Economic History Review* 35 (1982): 354–72.

3. References to tobacco or nicotine addiction are not common prior to 1900 (as revealed by an ngram search), though they are not entirely absent. John Quincy Adams in 1845, for example, in a letter to the Reverend Samuel H. Cox in Brooklyn wrote about how as a youth the future president had been "addicted to the use of tobacco in two of its mysteries, smoking and chewing." See Benjamin I. Lane, *The Mysteries of Tobacco* (New York: Wiley and Putnam 1846), 32. This is a favorite document of tobacco industry lawyers, introduced in trial after trial to postulate

the long-standing "common knowledge" of addiction (part of their "assumption of risk" defense).

4. Robert Heimann, *Tobacco and Americans* (New York: McGraw-Hill, 1960). In the seventeenth and eighteenth centuries paper-wrapped cigarettes were generally considered beggars' smokes. Cubans in the eighteenth century began using paper made from cotton for cigarette wrappers; see Susan Wagner, *Cigarette Country: Tobacco in American History and Politics* (New York: Praeger, 1971). In 1861, the Susini cigarette-rolling machine in Paris was said to have had a capacity of 3,600 cigarettes per hour; see Maurice Corina, *Trust in Tobacco* (London: Michael Joseph, 1975).

5. The best book on this is Marc Linder's encyclopedic (and underappreciated) *"Inherently Bad, and Bad Only": A History of State-Level Regulation of Cigarettes and Smoking in the United States Since the 1880s* (Iowa City: University of Iowa Faculty Books, 2012), http://ir.uiowa.edu/books/2/. The Florida State Board of Health in 1906 characterized the cigarette as "looked upon by all smokers as the very worst form of tobacco addiction" thanks to the filth contained therein (*Florida Health Notes*, July 1906, 10), but in 1909 this same publication reported with regard to cigarette use that "The general consensus is that moderation is not harmful and in some cases beneficial" (*Florida Health Notes*, August 1909, 119-20, http://www.archive.org/stream/annualreportstat1909flor/annualreportstat1909flor_djvu.txt).

6. See Tony Hyman, "Louis Susini's *La Honradez*: Cuban Cigarettes and the 1st Collectible: A National Cigar Museum Exhibit," http://cigarhistory.info/Cuba/Honradez.html. For samples of the images used on such wrappers, see http://hankwilliamslistings.com/ind-cuba.htm and http://cigarhistory.info/Cuba/Honradez_labels.html.

7. Luis Ceuvas-Alcober, "Die spanische 'Picadura,'" *Chronica Nicotiana* 4 (March 1943): 13.

8. Machines for labeling and boxing were also key; see Tony Hyman, "Louis Susini's La Honradez: Cuban Cigarettes & the 1st Collectible," http://www.cigarhistory.info/Cuba/Honradez.html.

9. Samuel Hazard, *Cuba a pluma y lápiz* (Cuba with pen and pencil), trans. Tony Hyman (Hartford, CT: Hartford Pub. Co., 1871).

10. Count Joseph de Susini-Ruiseco produced the first known estimates for total global cigarette consumption in his *La cigarette: Sa consommation et sa fabrication mécanique au moyen des machines système Susini* (Paris: Imprimerie A.-E. Rochette, 1872), figuring 294 billion cigarettes smoked per year. This is probably

a vast overestimate, perhaps even by an order of magnitude. Susini-Ruiseco also presented the first known calculations of the time it took to perform the different motions involved in hand-rolling cigarettes: 3 seconds to grasp and cut the paper, 10 seconds to pick up, spread, and shape the tobacco, 2 seconds to glue the paper, 3 to roll and seal the cigarette, and 2 to trim and finish the ends. He also figured that the maximum capacity of such a worker would be about 1,800 cigarettes per day, assuming a 10-hour working day with no breaks. Susini-Ruiseco was *administrateur directeur* of the Compagnie française des Tabacs, but it was actually Eugene Durand, a Paris engineer, who invented the machine that bears Susini's name. A number of Susini-Ruiseco's later patents can be found by searching Google Patents.

11. "Immense services au point de vue philanthropique" (Susini-Ruiseco, *La cigarette*, 10).

12. Robert N. Proctor, *Golden Holocaust: Origins of the Cigarette Catastrophe and the Case for Abolition* (Berkeley: University of California Press, 2011), 37–38.

13. Allan Brandt, *The Cigarette Century: The Rise, Fall, and Deadly Persistence of the Product that Defined America* (New York: Basic, 2007).

14. Richard B. Tennant, *The American Cigarette Industry: A Study in Economic Analysis and Public Policy* (New Haven, CT: Yale University Press, 1950), 69.

15. Continuous rod-making machines were later adapted for manufacture of inexpensive cigars and cigarillos—tube-shaped smoking devices that were sufficiently uniform to allow cost-effective machine manufacture. Cigar rolling had always been difficult to mechanize, given the difficulty of wrapping a full tobacco leaf; some advances on this had been made by the 1930s, and by the mid-1960s most American cigarillos were made on AMF's 270 machine, an adaptation for making full-size cigars from natural leaf capable of turning out sixteen cigars per minute. The 1960s saw a big push for higher-speed cigar making machines, using, for example, Arenco PMB's Bunchmaker SW/WV (made in Holland), which formed cigars using continuous rod making technology. This machine could churn out sixty to eighty shaped, double-length bunches (cigars without wrapping) per minute, with a maximum length of about ten inches. Mechanization of cigarette production was accompanied by a move to make cigars more similar to cigarettes—and the smoke pH of many was eventually lowered, to the point that most (cheap) cigars sold today are really just big brown cigarettes.

16. James A. Bonsack, "Cigarette-Machine," US Patent 238,640 (filed Sept. 4, 1880, and issued March 8, 1881); also his US Patent 247,795 (filed June 21, 1881,

and issued Oct. 4, 1881); Nannie M. Tilley, *The Bright-Tobacco Industry, 1860– 1929* (New York: Arno Press, 1972), 12-34.

17. James Bonsack was not the first to attempt a cigarette making machine using continuous process methods: Tennant reports that Albert H. Hook's in 1872 was the "first known attempt to manufacture a rolled and wrapped cigarette of indefinite length" (Tennant, *The American Cigarette Industry*, 17).

18. Vello Norman (Lorillard), "The History of Cigarettes," May 4, 1983, http:// legacy.library.ucsf.edu/tid/miy30e00.

19. Richard Kluger, *Ashes to Ashes: America's Hundred-Year Cigarette War, the Public Health, and the Unabashed Triumph of Philip Morris* (New York: Knopf, 1997); Brandt, *The Cigarette Century*.

20. Molins in 2000 acquired Filtrona Instruments and Automation, a maker of quality control instruments for the cigarette industry, which it renamed Cerulean. In 2002 the company acquired Arista Laboratories, a smoke constituent analytical facility in Richmond; see http://www.molins.com/company-history.aspx.

21. Robert N. Proctor, "The History of the Discovery of the Cigarette–Lung Cancer Link: Evidentiary Traditions, Corporate Denial, Global Toll," *Tobacco Control* 21 (2012): 87-91.

22. R. J. Reynolds in 1979, for example, was using a Molins Mark 9 cigarette making machine joined with a Hauni Max-S filter attacher to produce 5,600 cigarettes per minute; these cigarettes were then packed using a modified AMF high-speed packer to fill 250 packs per minute. The company was also using Sasib, Molins, and GDX-1 packers; see "Status of Manufacturing Department Projects and Activities for the Month of February 1979," March 14, 1979, http://legacy.library .ucsf.edu/tid/sok81b00.

23. Philip Morris, "Business Planning and Analysis," February 1987, http:// legacy.library.ucsf.edu/tid/wwk34e00. Philip Morris internal documents proposed long term goals for CPLH; see, for example, "Philip Morris Manufacturing Eight-Year Plan, 1980-1987," http://legacy.library.ucsf.edu/tid/wis95e00, C-3. For a great chart on Philip Morris's five-year plan for CPLH, see "Five Year Plan 1982-1986," http://legacy.library.ucsf.edu/tid/vpx83e00, and further discussion at http://legacy.library.ucsf.edu/tid/dkh07e00.

24. "RJR 1990 CPLH Estimate," May 9, 1991, http://legacy.library.ucsf.edu /tid/sje88h00; "Philip Morris U.S.A. Physical Parameters," March 2000, http:// legacy.library.ucsf.edu/tid/kbm75c00.

25. This machine was released in January 1992 at the Internepcon Japan '92

business exhibition; see Japan Tobacco Inc., "Annual Report 1992," http://legacy .library.ucsf.edu/tid/bsa30a00.

26. Edison's letter to Henry Ford is reprinted in Ford's *The Case Against the Little White Slaver* (Detroit: Henry Ford, 1916), 2.

27. Proctor, *Golden Holocaust*, 210–23.

28. Ibid., 31–35. Flue curing also had the advantage of producing a "cleaner" leaf, since heat was applied indirectly, preventing smoke or fumes from contaminating the tobacco. Flue curing by virtue of confining the flame also helped to prevent fires in barns. The first flues for conducting heat in barns date from around 1810, but the combination with charcoal was not made until the 1830s and broad commercialization did not come until after the Civil War, partly from the expense of outfitting barns with metal flues. The increasing availability of low-cost sheet metal after the 1860s dropped the price of the metal conduits used to transfer the heat, which also helped popularize the use of cast iron potbellied stoves, one of the first broadly distributed consumer appliances (from the 1830s). See Barry Donaldson and Bernard Nagengast, *Heat & Cold: Mastering the Great Indoors* (Ann Arbor, MI: ASHRAE, 1995).

29. See the Stanford Research into the Impact of Tobacco Advertising (SRITA) website at http://tobacco.stanford.edu/tobacco_main/slogans.php.

30. Brandt, *The Cigarette Century*; Kluger, *Ashes to Ashes*.

31. Brandt, *The Cigarette Century*; Proctor, *Golden Holocaust*; Naomi Oreskes and Erik M. Conway, *Merchants of Doubt* (New York: Bloomsbury, 2010); Louis M. Kyriakoudes, *Why We Smoke: History, Health, Culture, and the North American Origins of the Global Tobacco Epidemic* (forthcoming).

32. William L. Dunn, "Motives and Incentives in Cigarette Smoking," speech presented at CORESTA conference, Williamsburg, VA, October 22–28, 1972, http://legacy.library.ucsf.edu/tid/jfw56b00, 5.

33. Proctor, *Golden Holocaust*, 340–89.

34. Claude E. Teague Jr., "Research Planning Memorandum on the Nature of the Tobacco Business and the Crucial Role of Nicotine Therein," April 14, 1972, http://legacy.library.ucsf.edu/tid/kfp76b00.

35. Dunn, "Motives and Incentives in Cigarette Smoking."

36. Oscar Wilde's quote is from his 1891 *Picture of Dorian Gray* (New York: War, Lock and Co.), 65; the BAT quote is from Colin C. Greig, "Structured Creativity Group," British American Tobacco R&D, Southampton, 1984, http://legacy.library.ucsf.edu/tid/fsm86a99, 10.

37. U.S. Department of Health and Human Services, *Nicotine Addiction: A Report of the Surgeon General* (Washington DC: U.S. Government Printing Office, 1988), vi, identifies nicotine as "addicting in the same sense as are drugs such as heroin and cocaine." Compare also Lennox M. Johnston, "Tobacco Smoking and Nicotine," *Lancet* 243 (1942): 742.

38. This was one of the great fears of the industry from the point of view of litigation, that smoking might come to be widely recognized as addictive. Or as one Tobacco Institute memo put it: "Shook, Hardy reminds us, I'm told, that the entire matter of addiction is the most potent weapon a prosecuting attorney can have in a lung cancer/cigarette case. We can't defend continued smoking as 'free choice' if the person was 'addicted.'" See Paul Knopick to William Kloepfer, Sept. 9, 1980, http://legacy.library.ucsf.edu/tid/gkx74e00.

39. Roper Research Associates (for Philip Morris), "A Study of Cigarette Smokers' Habits and Attitudes in 1970," May 1970, http://legacy.library.ucsf.edu/tid/jyx81a00, 13, 18, 39.

40. Cigars are interestingly different from cigarettes in this respect; cigars are more often smoked more episodically and festively than cigarettes and are significantly less addictive and less likely to cause lung cancer. That is mainly because the smoke from traditional cigars is less often inhaled—the smoke is simply too harsh, too alkaline. In recent years, however, many manufacturers of cheaper varieties of cigars have manipulated the tobaccos used in those products to make the resulting smoke milder, less harsh, more inhalable. Today, most of the (cheaper) cigars sold in the United States are really more like big brown cigarettes—delivering an inhalable smoke with a pH less than 7. Fine, large, specialty cigars are somewhat more "honest" in this respect: they cause less harm because they have not been designed in such a way as to deliver inhalable smoke. Cigar makers did not participate directly in the denialist conspiracy and have never been vociferous in denying the health harms from smoke. Cigar makers have also been less deceptive in the basic design of their product: no cigar has ever had a filter, for example. Smokers of "filtered" cigarettes might want to reflect on this fact, that if filters confer some type of advantage, then why don't cigars ever have them? The fact is that the "filters" on cigarettes are not really filters; they don't keep smoke out of the smoker's mouth or lungs, and smokers of cigarettes with "filters" are just as likely to get cancer as smokers of "non-filtered" cigarettes; see Proctor, *Golden Holocaust*, 340–89.

41. Robert L. Bexon to ITL President Wilmat Tennyson and W. Sanders, File Viking, 1985, CTRL No. 3784, p. 2 (original emphasis), Database of Plaintiffs'

Counsels, Quebec Tobacco Litigation, https://tobacco.asp.visard.ca/GEIDEFile /iTL00684941.pdf?Archive=131232495941&File=Document.

## Chapter 4

1. Roland Barthes, "Toward a Psychosociology of Contemporary Food Consumption," in *Food and Culture: A Reader*, ed. Carole Counihan and Penny Van Esterik (London: Routledge, 2008); Massimo Montanari, *Culture of Food* (Cambridge: Blackwell, 1994), 63; Kathleen LeBesco and Peter Naccarato, *Edible Ideologies: Representing Food and Meaning* (Albany, NY: SUNY Press, 2008).

2. "By 2606, the US Diet Will Be 100% Sugar," *Whole Health Source* (blog), February 12, 2012, http://wholehealthsource.blogspot.com/2012/02/by-2606-us -diet-will-be-100-percent.html.

3. Montanari, *Culture of Food*, 95, 104–5, 118–21, 152, 167.

4. Hans Teuteberg, "The Birth of the Modern Consumer Age," in *Food, The History of Taste*, ed. Paul Freedman (Berkeley: University of California Press, 2007).

5. Sidney Mintz, *Sweetness and Power: The Place of Sugar in Modern History* (New York: Penguin, 1985), 15; Philip Gott and L. F. Van Houten, *All About Candy and Chocolate* (Chicago: National Confectioners' Association, 1958), 1–23; Tim Richardson, *Sweets: A History of Candy* (London: Bloomsbury, 2002), 43–57.

6. Robert Lustig, *Fat Chance: Beating the Odds against Sugar, Processed Food, Obesity & Disease* (New York: Hudson Street Press, 2012); L. M. Beidler, "The Biological and Cultural Role of Sweeteners," in *Sweeteners: Issues and Uncertainties*, ed. Academic Forum (Washington DC: National Academy of Science, 1975), 11–18.

7. Kevin Drum, "More on the Sugar Lobby," *Mother Jones*, February 7, 2010, http://www.motherjones.com/kevin-drum/2010/02/more-sugar-lobby.

8. L. A. G. Strong, *The Story of Sugar* (London: Weidenfeld & Nicolson, 1954), 38, 40, 42; Richardson, *Sweets*, 74–75.

9. Strong, *The Story of Sugar*, 50–51, 67–68; C. Trevor Williams, *Chocolate and Confectionery* (London: Leonard Hill, 1964), 1–4; Richardson, *Sweets*, 116.

10. Sidney Mintz, *Tasting Food, Tasting Freedom* (Boston: Beacon, 1996), 51–59.

11. Richardson, *Sweets*, 59, 63, 70–73; Strong, *The Story of Sugar*, 134–38; Williams, *Chocolate and Confectionery*, 12–15.

12. A. C. Hannah, *The International Sugar Trade* (Cambridge: Woodhead, 1996), chap. 2; William Dufty, *Sugar Blues* (Radnor, PA: Chilton, 1975), 14–21; Wendy Woloson, *Refined Tastes: Sugar, Confectionery, and Consumers in Nineteenth-Century America* (Baltimore, MD: The Johns Hopkins University Press, 2002), 5, 29–30.

13. John Rodrigue, *Reconstruction in the Cane Fields: From Slavery to Free Labor in Louisiana's Sugar Parishes, 1862–1880* (Baton Rouge, LA: LSU Press, 2001); Keith Sandiford, *The Cultural Politics of Sugar: Caribbean Slavery and Narratives of Colonialism* (New York: Cambridge University Press, 2000).

14. William Cronon, *Nature's Metropolis: Chicago and the Great West* (New York: Norton, 1991).

15. Michael Pollan, *The Omnivore's Dilemma* (New York: Penguin, 2006), 60, 85–89.

16. Lustig, *Fat Chance*. The best film on this is *King Corn*, directed by Aaron Woolf (Balcony Releasing, 2007).

17. Mintz, *Sweetness and Power*, 9, 13; Mintz, *Tasting Food, Tasting Freedom*, 19.

18. Mark M. Smith, *Sensing the Past: Seeing, Hearing, Smelling, Tasting and Touching in History* (Berkeley: University of California Press, 2007), 80–84; Mintz, *Sweetness and Power*, chap. 1; Harvey Levenstein, *Revolution at the Table: The Transformation of the American Diet* (New York: Oxford University Press, 1988), 31–33.

19. Mintz, *Sweetness and Power*, 138–41, 143, 187–88, 192–93, 198; Mintz, *Tasting Food, Tasting Freedom*, 13, 25–27.

20. Henry Weatherley, *Treatise on the Art of Boiling Sugar* (Philadelphia: Baird, 1865), 1–5; William Jeanes, *Modern Confectionery* (London: John Holt, 1864), 17–19, 23, 48, 53, 150–58, 182; H. Brauner, *H. Brauner's Instruction in Scientific Candy Making* (San Bernardino, CA: privately printed, 1915), 1–22; William Rigby, *Rigby's Reliable Candy Teacher* (New York: privately printed, 1902), 1–11; Charles Huling, *Notes on American Confectionery* (Philadelphia: privately printed, 1891), 15–18.

21. Jeanes, *Modern Confectionery*, 53; Brauner, *H. Brauner's Instruction in Scientific Candy Making*, 22; Victor Porter, *Practical Candy Making: Delicious Candies for Home or Shop* (New York: Frederick A. Stokes, 1929); George Herter and Russell Hofmeister, *History and Secrets of Professional Candy Making* (Waseca, MN: Herter's Inc., 1964), 6–11.

22. Gott and Van Houten, *All About Candy*, 1–23; Necco Company, *A Century of Candy Making* (Boston: Necco, 1947), 9–10, 20–21; National Equipment Company, *Candy and Chocolate Making Machinery* (Springfield MA: NEC, 1912), 1, 2, 39.

23. Michael Redclift, *Chewing Gum: The Fortunes of Taste* (New York: Routledge, 2004), 31.

24. Gum ads, *Collier's*, January 1, 1910, 17; July 8, 1911, 4; *Harper's Weekly*, November 25, 1899, 1221; *Youth's Companion*, March 14, 1912, back cover; May 18, 1905, back cover; August 8, 1907, 368.

25. Waverley Root and Richard de Rochemont, *Eating in America, A History* (New York: Morrow, 1976), 43–45; Redclift, *Chewing Gum*, 39–41.

26. Redclift, *Chewing Gum*, 35, 149–50.

27. Carolyn Wyman, *Jell-O, A Biography* (New York: Harcourt, 2001), ix–xi, 5, 14, 22.

28. Jell-O ads: *Ladies' Home Journal*, October 1903, 53; June 1910, 5; March 1916, 66.

29. Richardson, *Sweets*, 210; Woloson, *Refined Tastes*, 11–12, 15; Candy ads: *Youth's Companion*, February 9, 1899, 22; February 2, 1903, 71; January 17, 1907, 26; *Collier's*, March 6, 1909, back cover; May 15, 1915, 33; see also Cele Otnes and Elizabeth Pleck, *Cinderella Dreams: The Allure of the Lavish Wedding* (Berkeley: University of California Press, 2003), chap. 2; Elizabeth Pleck, *Celebrating the Family: Ethnicity, Consumer Culture, and Family Rituals* (Cambridge, MA: Harvard University Press, 2000), chap. 7.

30. Marion Nestle and Malden Nesheim, *Why Calories Count: From Science to Politics* (Berkeley: University of California Press, 2012).

31. Woloson, *Refined Tastes*, 26–49.

32. Ibid., 26–49, 52–3; Henry Bunting, *Specialty Advertising: The New Way to Build Business* (Chicago: Novelty News Press, 1910), 76. "Penny Candies Up to Date," *Confectioner's and Baker's Gazette*, November 10, 1905, 13–14; candy ad, *Youth's Companion*, October 27, 1892, 563; Necco, *A Century of Candy*, 20–28; Trade Cards and Label Collection, Col. 9 Box 2, Winterthur Museum Library, Winterthur, Delaware; Ray Broekel, *The Great American Candy Bar Book* (Boston: Houghton Mifflin, 1982), 49–50, 68–71, 112–14.

33. "The Confectionary," *Friend* 8 (1834): 141, cited in Woloson, *Refined Tastes*, 54 (see also 60–65); Sarah Rorer, "Why Sweets Are Not Good for Children," *Ladies' Home Journal*, March 1906, 38, 144; David Nasaw, *Children of the*

*City: At Work and at Play* (New York: Anchor, 2012), 118, 131; Patrick Porter, "Advertising in the Early Cigarette Industry: W. Duke, Sons & Company of Durham," *North Carolina Historical Review* 48 (1971): 35.

34. Google's ngram viewer is available online at http://books.google.com /ngrams.

35. On sugar in tobacco, see Robert N. Proctor, *Golden Holocaust: Origins of the Cigarette Catastrophe and the Case for Abolition* (Berkeley: University of California Press, 2011), 32–33.

36. William G. Clarence-Smith, *Cocoa and Chocolate, 1765–1914* (London: Routledge, 2000), 10–11; Sophie Coe and Michael Coe, *The True History of Chocolate* (London: Thames and Hudson, 1996), 31.

37. Coe and Coe, *The True History of Chocolate*, 13, 30–32; Mort Rosenblum, *Chocolate: A Bittersweet Saga of Dark and Light* (New York: North Point Press, 2005), 1–21, 51; J. C. Motamayor et al., "Cacao Domestication I: The Origin of the Cacao Cultivated by the Mayas," *Heredity*, 89 (2002): 380–86.

38. Girolamo Benzoni, *History of the New World*, cited in Coe and Coe, *The True History of Chocolate*, 108, also, 123, 208–9, 241. By 1860 few believed in the medical advantages of chocolate; see Clarence-Smith, *Cocoa and Chocolate*, 11–20; Woloson, *Refined Tastes*, 110–15.

39. Coffee didn't trickle down to the masses in Britain because Arabian (and later non-English colonial) coffee was more expensive than Indian tea — especially after the tea tax was removed in 1784; see Brian Cowan, *The Social Life of Coffee* (New Haven, CT: Yale University Press, 2005), 20, 22, 43, 77, 80; Tom Standage, *A History of the World in 6 Glasses* (New York: Walker, 2005), chaps. 7 and 8.

40. The comparison is more amusing in the original German, where *Sätze* plays on the double meaning of "coffee grounds" and "theorem": "Ein Mathematiker ist eine Maschine, die Kaffee in Sätze verwandelt."

41. A. Hausner, *The Manufacture of Preserved Foods and Sweetmeats* (London: Scott, Greenwood, 1912), 170–93; Gott and van Houton, *All about Candy*, 79; Michael Lasky, *The Complete Junk Food Book* (New York: McGraw-Hill, 1977), 46–47; Richardson, *Sweets*, 226; Clarence-Smith, *Cocoa and Chocolate*; and Coe and Coe, *The True History of Chocolate*, 241–42.

42. Richardson, *Sweets*, 228–29; Coe and Coe, *The True History of Chocolate*, 250, 116–17, 177–78; Woloson, *Refined Tastes*, 258–59; Clarence-Smith, *Cocoa and Chocolate*, 23–27; Baker and Co., *Cocoa and Chocolate* (Dorchester, MA: Baker and Co., 1899), 5–8.

43. Ray Broekel, *The Chocolate Chronicles* (Lombard, IL: Wallace-Homestead, 1985), 110–15; Woloson, *Refined Tastes*, 150–52. For analysis of recent uses of gender and the exotic in chocolate ads, see Ellen Moore, "Raising the Bar: The Complicated Consumption of Chocolate," in *Food for Thought: Essays on Eating and Culture*, ed. Lawrence Rubin (Jefferson, NC: McFarland, 2008).

44. Woloson, *Refined Tastes*, 145; Timothy Erdman, "Hershey: Sweet Smell of Success," *American History Illustrated* 29 (March–April 1994): 68.

45. Coe and Coe, *The True History of Chocolate*, 252–54.

46. Broekel, *The Chocolate Chronicles*, 22, 31, 34; Richardson, *Sweets*, 270–76; Michael D'Antonio, *Hershey: Milton S. Hershey's Extraordinary Life of Wealth, Empire and Utopian Dreams* (New York: Simon and Shuster, 2006); Joel Brenner, *Emperors of Chocolate: Inside the Secret Worlds of Hershey and Mars* (New York: Random House, 1999); John McMahon, *Built on Chocolate* (Los Angeles: General Publishing, 1998).

47. Andrew Smith, *Peanuts: The Illustrious History of the Goober Pea* (Urbana: University of Illinois Press, 2002), 75.

48. Broekel, *The Chocolate Chronicles*, 75, 80, 83.

49. Broekel, *Great American Candy Bar Book*, 11–40.

50. Stephen Jay Gould, "Phyletic Size Decrease in Hershey Bars," in *Hen's Teeth and Horse's Toes* (New York: Norton, 1980), 313–19.

51. An ngram search reveals that the expression "empty calories" is rare until the late 1950s.

52. A "snack" is not just a substance but a behavior, a verb ("to snack"), which means that we also have to appreciate that *any* food can be snacked, at least in principle. Snacking has to do with how and when we eat. Meal-less (or continuous) eating has a history, as does the whole idea of "meals" itself, deriving from historical practices of work, structures of family routines, the scheduling of pauses during work, and not just access to or even pressures to consume certain kinds of foods. Crucial then may be the schedule of work, or the forces tempting consumption.

53. Woloson, *Refined Tastes*, 67–76; Root and de Rochemont, *Eating in America*, 427–29; Paul Dickson, *The Great American Ice Cream Book* (New York: Atheneum, 1972), 15–20; Anne Cooper Funderburg, *Chocolate, Strawberry, and Vanilla: A History of American Ice Cream* (Bowling Green, OH: Bowling Green State University Press, 1995), 3, 73–75; Richardson, *Sweets*, 195, 202; Root and de Rochemont, *Eating in America*, 429.

54. Anne Cooper Funderburg, *Sundae Best: A History of Soda Fountains*

(Bowling Green, OH: Bowling Green State University Press, 2002), 62; Woloson, *Refined Tastes*, 79–99.

55. Elizabeth David, *Harvest of the Cold Months — The Social History of Ice and Ices* (New York: Viking, 1995).

56. Mark Lender and James Martin, *Drinking in America: A History* (New York: Free Press, 1982), 4–16; John H. Brown, *Early American Beverages* (Rutland, VT: Tuttle Press, 1966), 21, 77; Funderburg, *Sundae Best*, 78–84.

57. Joseph Priestley, *Directions for Impregnating Water with Fixed Air* (London: Johnson, 1772).

58. Funderburg, *Chocolate, Strawberry, and Vanilla*, 6–7; John Riley, *A History of the American Soft Drink Industry: Bottled Carbonated Beverages, 1807–1957* (Washington DC: American Bottlers of Carbonated Beverages, 1958), 23, 34; Stephen N. Tchudi, *Soda Poppery* (New York: Scribners, 1986), 8.

59. Riley, *A History of the American Soft Drink Industry*, 4–26, 48–63, 248–51.

60. Ibid., 4–10, 114.

61. Methyl anthranilate is a synthetic with the flavor and aroma of grapes, used in Kool-Aid.

62. Riley, *A History of the American Soft Drink Industry*, 4–10, 114; Beverage World, *Beverage World: A 100 Year History 1882–1982* (Great Neck, NY: Beverage World, 1982), 1–3, 218; Jasper Woodroof and G. Frank Phillips, *Beverages: Carbonated and Noncarbonated* (Westport, CT: Avi Publishing, 1974); "New Names for Soda Beverages," *Pharmaceutical Era*, June 11, 1896, 747.

63. Funderburg, *Sundae Best*, 44–49; "Some Soda Fountain Statistics," *Scientific American*, August 12, 1899, 99; James Tufts, *Arctic Soda Water Apparatus* (Boston: Wilson and Son, 1890), 45–47.

64. Tufts, *Arctic Soda Water Apparatus*, 4–21, 47, 87; Root and de Rochemont, *Eating in America*, 419; Funderburg, *Sundae Best*, 10–40; Funderburg, *Chocolate, Strawberry, and Vanilla*, 97, 41–42; Riley, *A History of the American Soft Drink Industry*, 68, 90–91.

65. In order to seal the deal, many fountains dropped their wine-"flavored" sodas (to reinforce their claim of a new sort of refreshment). J. C. Furnas, *The Life and Times of the Late Demon Rum* (New York: Putnam, 1965), 55, 88–97, 122–34, 168–69, 284–85; Funderburg, *Sundae Best*, 88–92.

66. Funderburg, *Sundae Best*, 93–94, 201; "The Chas. E. Hires Co.," *Pharmaceutical Era*, June 15, 1893, 84; Tchudi, *Soda Poppery*, 21–23; Hire's ads: *Youth's*

*Companion*, April 11, 1901, 194; *Ladies' Home Journal*, May 1903, 35. Winterthur advertisements, Collection 214, Box 1 and Trade Cards and Labels Collection, Col. 9, Boxes 3 and 8, Winterthur Museum Library, which contains many examples of root beer and sarsaparilla products touted as health drinks.

67. Frank N. Potter, *The Book of Moxie* (Paducah, KY: Collector's Books, 1987), 162, 82–83, 88.

68. Jeffrey L. Rodengen, *The Legend of Dr. Pepper/Seven-Up* (Fort Lauderdale, FL: Write Stuff, 1995), 29–40; Funderburg, *Sundae Best*, 67–72.

69. Woodroof and Phillips, *Beverages*, 1–3; Clicquot Club ads, *Collier's*, February 18, 1914, 26; June 12, 1915, 39.

70. Michael Witzel and Gyvel Young-Witzel, *Soda Pop!* (Stillwater, MN: Voyageur, 1998), 75–76; Mark Pendergrast, *For God, Country and Coca-Cola: The Unauthorized History of the Great American Soft Drink and the Company that Makes It* (New York: Scribner's 1993), 22; J. C. Louis and Harvey Yazijian, *The Cola Wars* (New York: Everest House, 1980), 18; Funderburg, *Sundae Best*, 74–75.

71. Joseph F. Spillane, *Cocaine: From Medical Marvel to Modern Menace in the United States, 1884–1920* (Baltimore, MD: The Johns Hopkins University Press, 2002); Tim Madge, *White Mischief: A Cultural History of Cocaine* (Edinburgh: Mainstream Publishing, 2001).

72. Pendergrast, *For God, Country and Coca-Cola*, 30, 36–7.

73. Louis and Yazijian, *The Cola Wars*, 15–18, 76; Pat Watters, *Coca-Cola: An Illustrated History* (Garden City, NY: Doubleday, 1978), 13–14; "To Pause and Be Refreshed," *Fortune*, July 1931, 65, 111; Richard Tedlow, *New and Improved: The Story of Mass Marketing in America* (New York: Basic, 1990), 22–26; Root and de Rochemont, *Eating in America*, 422; Tchudi, *Soda Poppery*, 25–34.

74. Bob Stoddard, *Pepsi 100 Years* (Los Angeles: General Publishing, 1997), 11, 24, 34; Woodroof and Phillips, *Beverages*, 70–73.

75. Beverage World, *Beverage World*, 201; Tedlow, *New and Improved*, chap. 3.

76. Mintz, *Sweetness and Power*, 130–34.

77. Ibid., 201–3.

78. David M. Cutler, Edward L. Glaeser and Jesse M. Shapiro, "Why Have Americans Become More Obese?," *Journal of Economic Perspectives* 17, no. 3 (Summer 2003): 93. Katherine Flegal et al., "Prevalence of Obesity and Trends in the Distribution of Body Mass Index among US Adults, 1999-2010," *JAMA* 307 (2012): 491–97.

## Chapter 5

1. The full text of Hans Christian Andersen's *The Little Mermaid* (1836) can be found at http://hca.gilead.org.il/li_merma.html.

2. Oliver Read and Walter Welch, *From Tin Foil to Stereo: Evolution of the Phonograph* (Indianapolis: Bobbs-Merrill, 1976), 1–2; Jacques Attali, *Noise: The Political Economy of Music* (Minneapolis: University of Minnesota Press, 1985), 87.

3. K. LaGrandeur, "The Talking Brass Head as a Symbol of Dangerous Knowledge in *Friar Bacon* and in *Alphonsus, King of Aragon*," *English Studies* 5 (1999): 408–22; Cyrano de Bergerac, *Histoire comique en voyage dans la Lune* (1649; London: Doubleday, 1899); Alfred Mayer, "Edison's Talking Machine," *Popular Science Monthly*, April 1878, 719–20; David Lindsay, "Talking Head," *Invention and Technology*, Summer 1997, 56–63; F. Rabelais, *The Histories of Gargantua and Pantagruel* (Harmondsworth, UK: Penguin, 1985), fourth book, chaps. 55 and 56, 566; Charles Grivel, "The Phonograph's Horned Mouth," in *Wireless Imagination: Sound, Radio, and the Avant-Garde*, ed. Douglas Kahn and George Whitehead (Cambridge, MA: MIT Press, 1992), 43.

4. Evan Eisenberg, *The Recording Angel: Music, Records and Culture from Aristotle to Zappa*, 2nd ed. (New Haven, CT: Yale University Press, 2005), 12–14.

5. Thomas Edison, "The Perfected Phonograph," *North American Review*, June 1888, 642–43.

6. *Scientific American*, November 17, 1877, 304.

7. "The Talking Phonograph," *Scientific American*, December 22, 1877, 384; Roland Gelatt, *The Fabulous Phonograph, 1877–1977* (New York: Macmillan, 1977), 29; Count du Moncel, *The Telephone, Microphone and the Phonograph* (London: Kegan Paul, 1879), 306–26; Wyn Wachhorst, *Thomas Alva Edison: An American Myth* (Cambridge, MA: MIT Press, 1981), 20; Ronald Clark, *Edison, The Man who Made the Future* (New York: Putnam, 1977), 76; Myrna Frommer, "How Well Do Inventors Understand the Cultural Consequences of their Inventions?" (PhD diss., New York University, 1987), 67.

8. Gelatt, *The Fabulous Phonograph*, 32; Greg Milner, *Perfecting Sound Forever: An Aural History of Recorded Music* (New York: Faber and Faber, 2009), 23.

9. A. J. Millard, *America on Record* (New York: Cambridge University Press, 1995), 24–28.

10. "The Talking Phonograph," 384–85; "The Phonograph," *Harper's Weekly*,

March 30, 1878, 249–50; T. C. Fabrizio and George Paul, *Antique Phonograph Advertising* (Atglen, PA: Schiffer, 2002), vii; Frederick Garbit, *The Phonograph and its Inventor: Thomas Alva Edison* (Boston: Gunn Bliss, 1878), 8; "Phonograph," *Chicago Tribune*, March 16, 1878, 16.

11. "The Phonograph," *Harper's Weekly*, 249–50; Thomas Edison, "The Phonograph and its Future," *The North American Review*, May/June, 1878, 527–32.

12. Edison, "The Phonograph and Its Future," 527–36; Norman Lockyear, "The Phonograph," *Nature*, May 30, 1878, 117; William Tegg, *The Telephone and the Phonograph* (London: McCorquodale, 1878), 40; George Prescott, *The Speaking Telephone, Talking Phonograph and Other Novelties* (New York: Appleton, 1878), 306; Mary Collins, *Thomas Edison and Modern America: A Brief History with Documents* (New York: Macmillan, 2002), 65–72.

13. Edison, "The Phonograph and its Future," 531; "The Phonograph," *Harper's Weekly*, 249–50; Michael Chanan, *Repeated Takes: A Short History of Recording and its Effects on Music* (London: Verso, 1995), 3.

14. Edison, "The Perfected Phonograph," 648–49; David Suisman, *Selling Sounds: The Commercial Revolution in American Music* (Cambridge, MA: Harvard University Press, 2009), 95.

15. "Reproduction of Articulate Speech and Other Sounds," *Scientific American*, July 14, 1888, 15–16; "Progress of the New Edison Electric Phonograph," *Scientific American*, May 26, 1888, 320; "Wireman's Wrecklessness," *New York Times*, May 12, 1888, 8; Frommer, "How Well Do Inventors Understand the Cultural Consequences of their Inventions?," 74.

16. "Progress of the New Edison Electric Phonograph," 321; Gelatt, *The Fabulous Phonograph*, 32–36, 41–43; Millard, *America on Record*, 28–35, 38–41; Allen Koenigsberg, *Edison Cylinder Records, 1889–1912* (New York: Stellar Productions, 1969), xii.

17. *Phonogram*, April 1891, 12, and October 1891, 13; Read and Welch, *From Tin Foil to Stereo*, 292.

18. Millard, *America on Record*, 38–41; Read and Welch, *From Tin Foil to Stereo*, 105–7; Frommer, "How Well Do Inventors Understand the Cultural Consequences of their Inventions?," 77; A. O. Tate, *Edison's Open Door* (New York: Dutton, 1938), 253.

19. Read and Welch, *From Tin Foil to Stereo*, 115–118, 301–332; Frommer,

"How Well Do Inventors Understand the Cultural Consequences of their Inventions?," 78–80; Suisman, *Selling Sounds*, 94–100; Q. David Bowers, *Put Another Nickel In: A History of Coin-Operated Pianos and Orchestrions* (New York: Vestal, 1966).

20. Edison ad, *Phonogram, November 1892, 35*; Koenigsberg, *Edison Cylinder Records*, xviii.

21. "Phonograph Improved," *New York Times*, April 5, 1896, 16; "Improved Process of Duplicating Phonograph Records," *Scientific American*, April 20, 1901, 242; Koenigsberg, *Edison Cylinder Records*, xix; Gelatt, *The Fabulous Phonograph*, 56–57; James Weber, *The Talking Machine: The Advertising History of the Berliner Gramophone and Victor Talking Machine* (Midland, ON: Adio, 1997), 2–3; Millard, *America on Record*, 116; George Tewksbury, *Complete Manual of The Edison Phonograph* (Newark, NJ: U.S. Phonograph Company, 1897), 13–19, 27.

22. Fabrizio and Paul, *Antique Phonograph Advertising*, 47.

23. Weber, *Talking Machine*, 5.

24. Read and Welch, *From Tin Foil to Stereo*, 119–21; Gelatt, *The Fabulous Phonograph*, 62–64; Millard, *America on Record*, 45–47.

25. Frommer, "How Well Do Inventors Understand the Cultural Consequences of their Inventions?," 86.

26. Attali, *Noise*, 89.

27. Peter Copeland, *Sound Recordings* (London: British Library, 1991), 12–15; Gelatt, *The Fabulous Phonograph*, 46–55, 81; Timothy Day, *A Century of Recorded Music* (New Haven, CT: Yale University Press, 2000), 9; John Harvith and Susan Harvith, eds., *Edison, Musicians, and the Phonograph* (Westport, CT: Greenwood, 1987), 135.

28. Koenigsberg, *Edison Cylinder Records*, 53–54, 111–35; Gelatt, *The Fabulous Phonograph*, 75–80.

29. "A Talking Machine Fight," *New York Times*, November 20, 1898, 7; Gelatt, *The Fabulous Phonograph*, 110–13, 130–33; Millard, *America on Record*, 49–50; Chanan, *Repeated Takes*, 25; E. R. Fenimore Johnson, *His Master's Voice Was Eldridge R. Johnson* (Milford, DE: State Media, 1974), 54–55; Suisman, *Selling Sounds*, 102.

30. Millard, *America on Record*, 54–56, 76–77; Chanan, *Repeated Takes*, 54.

31. *Voice of the Victor*, September 1910, 5; *Voice of the Victor*, July 1909, 4; Read and Welch, *From Tin Foil to Stereo*, 68.

32. Victor ads featuring Nipper, *Cosmopolitan*, October 1901, 733; William Jenkins, *The Romance of Victor* (Camden, NJ: Victor, 1927); *Ladies' Home Journal*, February 1912, 78, and April 1912, 92; Weber, *Talking Machine*, 73–79; see also Arnold Schwartzman, *Phono-Graphics: The Visual Paraphernalia of the Talking Machine* (San Francisco: Chronicle Books, 1993), 13.

33. Edison ad, *Scientific American*, January 26, 1901, 62.

34. "A New Permanent Phonograph Record," *Scientific American*, March 9, 1901, 147; "Improved Process of Duplicating Phonograph Records," *Scientific American*, April 20, 1901, 242; Cylinder Preservation and Digitization Project, University of California Santa Barbara Library, http://cylinders.library.ucsb.edu /history-goldmoulded.php.

35. Edison ads: *Collier's*, November 28, 1908, 61, and March 19, 1910, 6; *Edison Phonographic Monthly*, October 1912, 3, and May 1914, 3; see also Collins, *Edison and Modern Americ*a, 78; Koenigsberg, *Edison Cylinder Records*, xxii–lvi; Fabrizio and Paul, *Antique Phonograph Advertising*, 50–51; Harvith and Harvith, *Edison, Musicians, and the Phonograph*, 7; Millard, *America on Record*, 66–68, 78–81; Read and Welch, *From Tin Foil to Stereo*, 165; Weber, *Talking Machine*, 94–95.

36. "Victor Disc Talking Machine" leaflet, ca. 1902, Hagley Museum Library, Greenville, DE; Victor ads: *Collier's* November 19, 1902, 73; *Collier's*, December 4, 1904, 2; *Voice of the Victor*, 1919, 5; Weber, *Talking Machine*, 112; "Victrola Newest Models, 1923," advertising leaflet, Hagley Museum Library.

37. Gelatt, *The Fabulous Phonograph*, 71; National Phonograph Company, "The Phonograph and How to Use it" (n.p., 1900), Hagley Museum Library; "Edison Diamond Disc Phonographs," np, 1914, Hagley Museum Library.

38. Victor Disk Talking Machine catalog, ca. 1900; Victor Taking Machine Co, 1910 catalog, Hagley Museum Library; Victor ads: *Outlook*, January 1, 1910, 2, and *Voice of the Victor*, 1919, 30; Wanamaker Co. "Great Phonographs Price List," ca. 1920, Hagley Museum Library; Weber, *Talking Machine*, 94–99; *Youth's Companion*, January 21, 1915, 40; *Voice of the Victor*, 1919, 42.

39. Columbia ads, *McClure's*, November 1907, 103, and January 1, 1909, 47.

40. Edison ads: *Edison Phonograph Monthly*, June 1903, 11; *Youth's Companion*, June 25, 1908, 312, and *Ladies' Home Journal*, November 1907, 84; see also Gelatt, *The Fabulous Phonograph*, 164–68.

41. Edison ads: *Collier's*, March 28, 1908, 4; *Edison Phonograph Monthly*,

April 1906, 7; May 1906, 9; March 1910, 22–23; October 1911, 8–9; *McClure's*, November 1907, 24; *Collier's*, June 20, 1909, 7, and July 31, 1909, 27; see also Read and Welch, *From Tin Foil to Stereo*, 194.

42. Harvith and Harvith, *Edison, Musicians, and the Phonograph*, 7, 9.

43. Edison ad, *McClure's*, November 1907, 399; Edison letter, February 8, 1915, cited in Collins, *Edison and Modern America*, 178–80.

44. Edison ads: *Edison Phonograph Monthly*, December, 1903, 12, 14; April 1904, 7; April 1905, 13; August 1911, 11.

45. "Musical Ideals of Thomas A. Edison," *Edison Phonograph Monthly*, January 1914, 1; Thomas Edison Inc., *Mood Music: A Compilation of 112 Edison Re-Creations According to "What They Will Do for You"* (Orange, NJ: Edison, 1921), 28–31; Milner, *Perfecting Sound Forever*, 47.

46. Fred Barnum, *A Century of Electronic Communications Milestones from Camden NJ, 1900–2001* (Camden, NJ: RCA, 2001), 46; *Voice of the Victor*, May 1907, 2; May 1908, 9; October 1911, cover; Victor ads: *Ladies' Home Journal*, July 1905, 2, and *McClure's*, February 1912, 22; see also Frommer, "How Well Do Inventors Understand the Cultural Consequences of their Inventions?," 89–90.

47. Read and Welch, *From Tin Foil to Stereo*, 154; Schwartzman, *Phono-Graphics*, 58; "Edison Diamond Disc Phonographs," (n.p., 1913), Hagley Museum Library; Suisman, *Selling Sounds*, 101–24 (quotation on 104), 125–49; see also Marsha Siefert, "The Audience at Home: The Early Recording Industry and the Marketing of Musical Taste," in *Audiencemaking: How the Media Create the Audience*, ed. James Ettema and D. Charles Whitney (Thousand Oaks, CA: Sage, 1994).

48. Victor ads: Schwartzman, *Phono-Graphics*, 42; Consolidated Talking Machine, "Improved Gramophone, 1903," Hagley Museum Library; Victor ads: *Ladies' Home Journal* Oct. 1913, 102; and *Youth's Companion*, February 10, 1916, 84.

49. Victor ads: *Outlook*, February 4, 1911, 2; *Collier's* October 10, 1908, back cover; *Voice of the Victor*, July 1909, 4–5; Fabrizio and Paul, *Antique Phonograph Advertising*, 8; see also Gelatt, *The Fabulous Phonograph*, 142–44.

50. Victrola XVI 1907 ad, Schwartzman, *Phono-Graphics*, 48; Victor Victrola 1910 leaflet; Victor/Victrola catalog, 1909; "Will there be a Victrola in your home this Christmas," 1915, Hagley Museum Library; Victor ads: *Voice of the* Victor, May 1907, 4; *Voice of the Victor*, May 1908, 4; July 1912 cover.

51. Millard, *America on Record*, 68–70; Oscar Saenger, "The Oscar Saenger

Course in Vocal Training: A Complete Course of Vocal Study of the Tenor Voice," in *Music, Sound, and Technology in America: A Documentary History of Early Phonograph, Cinema, and Radio*, ed. Timothy D. Taylor, Mark Katz, and Tony Grajeda (Durham, NC: Duke University Press, 1912), 103–4; Victor ad for Oscar Saenger records, *McClure's*, April 1917, 1; Victor ads: *Boston Globe*, April 9, 1918, 4; *Voice of the Victor*, November 1907, 8; October 1914, 18–19; *Ladies' Home Journal*, January 1912, 56; see also Jacob Smith, *Vocal Tracks: Performance and Sound Media* (Berkeley: University of California Press, 2008), chaps. 3 and 4.

52. Victor catalog, 1903, Hagley Museum Library; Victor ads: *Ladies' Home Journal*, October 1906, 79; Schwartzman, *Phono Graphics*, 38, 42; *Voice of the Victor*, May–June 1911, 1–4; *Voice of the Victor*, May 1908, 7; see also Victor Victrola catalog, 1910; Victor catalog, 1918, both in Hagley Museum Library.

53. Eldridge Johnson speech, *Voice of the Victor*, July 1909, 1.

54. Victor ads: *Talking Machine World*, July 1913, cover; *Voice of the Victor*, July 1907, 5.

55. *Voice of the Victor*, October 1911, 6; Victor ad, *McClure's* September 1905, 2.

56. Columbia Record catalog, 1923, Hagley Museum Library; Mark Coleman, *Playback: From the Victrola to MP3, 100 Years of Music, Machines, and Money* (Cambridge, MA: Da Capo, 2003), 20–27.

57. Columbia record ads, *Ladies' Home Journal*, April 1916, 65; June 1916, 51.

58. National Phonograph Co. catalog, 1899, 41, Hagley Museum Library.

59. Victor ads: Schwartzman, *Phonographics*, 36; *Ladies' Home Journal*, January 1912, 56, and February 1913, 72; see also Victor Talking Machine Co., "How to Get the Most Out of Your Victrola," 1918, Hagley Museum Library.

60. "How to Get the Most out of your Victrola"; National Phonograph Co. catalog, 1905, 1–5, Hagley Museum Library.

61. For further exploration of the electronic appliance's domestication of mass culture, see Lynn Spigel, *Make Room for TV: Television and the Family Ideal in Postwar America* (Chicago: University of Chicago Press, 1992), chap. 1; Cecelia Tichi, *Electronic Hearth: Creating an American Television Culture* (New York: Oxford University Press, 1991), 16, 19, 29, 32.

62. Katherine Grier, *Culture and Comfort: Parlor Making and Middle-Class Identity, 1850–1930* (Washington DC: Smithsonian Institution Press, 1998); Mihaly Csikszentmihalyi and Eugene Rochberg-Halton, *The Meaning of Things: Domestic Symbols and the Self* (New York: Cambridge University Press, 1981).

63. Edison Phonograph leaflet (ca. 1900) in Fabrizio and Paul, *Antique Phono-*

*graph Advertising*, 91; National Phonograph Company, 1906 catalog, Hagley Museum Library; "A Master Product of a Master Mind: New Edison Diamond Amberola, 1917-18" (Orange, NJ: Edison, 1918); 6-7.

64. Collins, *Edison and Modern America*, 138-140; Edison ad, *Youth's Companion* January 30, 1908, 55.

65. Edison ads, *Ladies' Home Journal*, January 1912, 33; *Collier's*, October 15, 1910, 13; and January 29, 1911, back cover.

66. Victor ad, *Ladies' Home Journal*, August 1913, 46.

67. Koenigsberg, *Edison Cylinder Records*, xxiv.

68. Gelatt, *The Fabulous Phonograph*, 178, 208-18; Eisenberg, *Recording Angel*, 18-22, 39; Suisman, *Selling Sounds*, chap. 6.

69. John Philip Sousa, "The Menace of Mechanical Music," *Appleton's Magazine*, 8 (1906): 278-84; Gelatt, *The Fabulous Phonograph*, 146-47, 190-91; Eisenberg, *Recording Angel*, 14-15, 144-47, 171.

70. Eisenberg, *Recording Angel*, 23, 38, 43; Chanan, *Repeated Takes*, 7, 9; F. Lesure and R. L. Smith, eds., *Debussy on Music* (New York: Knopf, 1977), 288; Day, *A Century of Recorded Music*, 213.

71. Chanan, *Repeated Takes*, 12-13, 19-20.

## Chapter 6

1. Donald Lowe, *History of Bourgeois Perception* (Chicago: University of Chicago Press, 1982), 13. Sources on the Victorian quest for reducing smell and noise include Alain Corbin, *The Foul and the Fragrant: Odor and the French Social Imagination* (Cambridge, MA: Harvard University Press, 1986), chap. 1; Robert Jütte, *A History of the Senses: From Antiquity to Cyberspace* (Cambridge: Polity, 2005), 181-85, 266-69, 272. 275-76; Mark M. Smith, *Sensing the Past* (Berkeley: University of California Press, 2007), 8-27, 59-74; Jim Drobnick, ed., *The Smell Culture Reader* (Oxford: Berg, 2006).

2. John Hammond, *The Camera Obscura* (Bristol, UK: Adam Hilger, 1981), 1-10; Laurent Mannoni, *The Great Art of Light and Shadow: Archaeology of the Cinema*, trans. and ed. by Richard Crangle (Exeter, UK: University of Exeter Press, 2000), 3-27.

3. Jonathan Crary, *Techniques of the Observer: On Vision and Modernity in the Nineteenth Century* (Cambridge, MA: Harvard University Press, 1990), 39.

4. Mary Marien, *Photography, A Cultural History* (London: Lawrence King, 2006), 4–6.

5. Charles Musser, *The Emergence of Cinema* (Berkeley: University of California Press, 1990), 24–25.

6. John Pepper, *Scientific Amusements for Young People* (London: Routledge, 1868), 67; Mannoni, *The Great Art of Light and Shadow*, 3–27, 77, 79, 93, 115–34, 136–75, 257; Musser, *The Emergence of Cinema*, 22–24.

7. T. C. Hepworth, *Book of the Lantern* (New York: Edward Wilson, 1889), 14–15; David Robinson, "Magic Lantern Shows," *Encyclopedia of Early Cinema*, ed. Richard Able (New York: Routledge, 2005), 409–8.

8. Musser, *The Emergence of Cinema*, 30–31 35–37; T. Milligan, *Illustrated Catalogue of Magic Lantern Apparatus* (Philadelphia, 1882); Thomas Hall, *Hall's Illustrated Catalogue of Magic Lanterns* (Boston, 1881); Jesse Cheyney, *Catalogue of Magic Lanterns, Stereopticons, and Views* (New York, 1876).

9. Stephan Oettermann, *A Panorama History of a Mass Medium* (New York: Zone, 1997), 7, 11–15, 20–22.

10. Germain Bapst, *Essai sur l'histoire des panoramas* (Paris, 1889), 19; Oettermann, *A Panorama History*, 31–33; Bernard Comment, *The Painted Panorama* (New York: Harry Abrams, 1999), 27–30, 57–64.

11. Oettermann, *A Panorama History*, 314, 323–25; John Banvard, *Panorama of the Mississippi River, Painted on 3 Miles of Canvas* (Boston: Putnam, 1847), 5, 44; William Burr, *Burr's Pictorial Voyage to Canada, American Frontier and the Saguenay* (Boston: Dutton, 1850), 43, 45 (quotation on 4); John F. McDermott, *Lost Panoramas of the Mississippi* (Chicago: University of Chicago Press, 1971), vii, 5–8; Joseph Arrington, "Henry Lewis' Moving Panorama of the Mississippi River," *Louisiana History* 6, no. 3 (Summer 1965): 239–72.

12. Oettermann, *A Panorama History*, 80.

13. Beaumont Newhall, *The Daguerreotype in America* (New York: Duell, Sloan & Pearce, 1961), 9–11.

14. Comment, *The Painted Panorama*, 57–60, 83, 133 (quotation on 13).

15. Mannoni, *The Great Art of Light and Shadow*, xxv, 28.

16. In Britain, an inventor by the name of William H. Fox Talbot developed yet another process, the calotype, using silver halide on paper to create a negative from which multiple (though blurred) copies could be made. Hammond, *The Camera Obscura*, 9–18, 48, 73, 104; Newhall, *The Daguerreotype in America*, 12–13;

Pierre Bourdieu, *Photography: A Middle-Brow Art* (Stanford, CA: Stanford University Press, 1990), 195; Reese Jenkins, *Images and Enterprise: Technology and the American Photographic Industry, 1839 to 1925* (Baltimore, MD: The Johns Hopkins University Press, 1975), 31.

17. Jenkins, *Images and Enterprise*, 10–11, 16, 19; Newhall, *The Daguerreotype in America*, 20.

18. Jenkins, *Images and Enterprise*, 3–5.

19. Ibid., 38–45, 48–49; William Darrah, *The World of Stereographs* (London: Yacht Press, 1997), 24–29.

20. Jenkins, *Images and Enterprise*, 85–86.

21. Ibid., 50; Crary, *Techniques of the Observer*, 128; Thomas Hawkins, *Instruments of the Imagination* (Princeton, NJ: Princeton University Press, 1995), Marien, *Photography, A Cultural History*, 82–3; Darrah, *World of Stereographs*, 4, 10–11, 21; Edward Earle, *Points of View: The Stereograph in America—A Cultural History* (Rochester, NY: University of Rochester Press, 1979), 2–3, 18–19, 30, 32, 60, 64.

22. Marien, *Photography, A Cultural History*, 23, 51–56, 63, 77, 79; Martin Jay, "Scoptic Regimes of Modernity," in *Visual Culture Reader*, ed. Nicholas Mirzoeff (London: Routledge, 1998), 66–69.

23. Oliver Holmes, "The Stereoscope and the Stereograph," *Atlantic Monthly*, June 1859, 744 (original emphasis); Oliver Holmes, "Stereoscope: or Travel Made Easy," *The Athenaeum*, March 20, 1858, 371.

24. Earle, *Points of View*, 12; Holmes, "The Stereoscope and Stereograph," 744.

25. For a full account of these changes, see Jenkins, *Images and Enterprise*, 67–76, 81–84.

26. Ibid., 98–100, 112, 114–15, 123, 181.

27. Eastman Film Co., *Kodak Primer* (Rochester, NY, 1888).

28. Nancy West, *Kodak and the Lens of Nostalgia* (Charlottesville: University Press of Virginia, 2000), 30–31, 49, 63. See also Colin Ford and Karl Steinworth, eds., *You Press the Button, We Do the Rest: The Birth of Snapshot Photography* (London: D. Nishen, in association with the National Museum of Photography, Film and Television, 1988).

29. West, *Kodak and the Lens of Nostalgia*, 20–25, 160; "Make your Kodak Autographic," amateur ads, 1913–1917, #2; "The Picture Worth Taking is Worth

Keeping," Kodak Ad Collection, 1926–27, #11; all in Eastman House Archives, Rochester, NY.

30. Gary Cross, *All-Consuming Century* (New York: Columbia University Press, 2000), chaps. 2 and 3.

31. West, *Kodak and the Lens of Nostalgia*, 20.

32. Ibid., 1–5, 14. On collecting, see Susan Stewart, *On Longing: Narratives of the Miniature, the Gigantic, the Souvenir, the Collection* (Durham, NC: Duke University Press, 1993), esp. 151–69; Russell Belk, *Collecting in a Consumer Society* (London: Routledge, 1995), 65–104; Jean Baudrillard, *The System of Objects* (1968; London: Verso, 1996), 85–108; John Potvin and Alla Myzelev, eds., *Material Cultures, 1740–1920* (Burlington, VT: Ashgate, 2009).

33. West, *Kodak and the Lens of Nostalgia*, 139.

34. Ibid., 73–90; Kodak ad: *Ladies' Home Journal*, June 1903, back cover.

35. Beth Bailey, *From Front Porch to Back Seat: Courtship in 20th Century America* (Baltimore, MD: The Johns Hopkins University Press, 1988).

36. Kodak ads: untitled, Amateur and Professional Copy, 1895-1907 #6, n.d.; "Kodak Brings Your Vacation Home," Amateur and Professional Copy, 1926–30, July 1930, #5; *Kodakery*, January 1918, 15, all in Eastman House Archive.

37. Ad no. 5293, Kodak Ad Collection, 1926–27; *Kodakery*, October 1913, 1, and November 1917, 16–17, all in Eastman House Archives.

38. West, *Kodak and the Lens of Nostalgia*, 11–12; Gary Cross, *The Cute and the Cool* (New York: Oxford University Press, 2004), chap. 3; Lindsay Smith, *The Politics of Focus: Women, Children and Nineteenth-Century Photography* (Manchester: Manchester University Press, 1998).

39. Brownie images are in *Youth's Companion*, April, 4, 1900, 22, 4; August 2, 1900, 381; November 28, 1901, back cover. See also Marc Oliver, "George Eastman's Modern Stone-Age Family: Snapshot Photography and the Brownie," *Technology and Culture* 48, no. 1 ( January 2007): 1–20.

40. Kodak ads: "Capturing Sweet Sixteen," Kodak Ad Collection, Amateur and Professional Copy, 1911-1914, 1911, #1; "Pride of Firsts," General Ads, 1928–29, 1928, #2; "Days that Would be Gone Forever," General Ads, 1926–1930, May 1928; "They are Boys and Girls so Short a Time," Amateur and Professional Copy, 1926–30, October 1929, #5; "There was Grandma," Amateur and Professional Copy, 1926–30, October 1928, #4; "In the Years to Come," General Advertising Copy 1928–29, July 1928, #3; all in Eastman House Archives.

41. Roland Barthes, *Camera Lucida: Reflections on Photography* (New York: Hill and Wang, 1982), 4.

42. West, *Kodak and the Lens of Nostalgia*, 139, 154.

43. Susan Sontag, *On Photography* (New York: Anchor, 1978), 24, 34–41.

44. Mannoni, *The Great Art of Light and Shadow*, 324, 340–41, 350; Mina Hammer, *History of the Kodak* (New York: House of Little Books, 1940), 10–18.

45. Musser, *The Emergence of Cinema*, 50–51, 62; *A Treasury of Early Cinema*, UCLA Film and Television Archives (hereafter cited as FTA), DVD 30; *Landmarks of Early Film*, FTA, DVD 27.

46. Paolo Cherchi Usai, "Eastman Kodak Company," in Able, *Encyclopedia of Early Cinema*, 197–98.

47. Movie cameras required the intermittent movement of the film across the lens to capture a series of images. This meant translating the rotary motion of a crank, spring winding, or electric motor into a discontinuous motion at the aperture of a camera or projector. Edison's kinetographic camera used two toothed discs to this effect. Charles Musser, "Kinetoscope," in Able, *Encyclopedia of Early Cinema*, 358–59; W. K. L. Dickson and Antonia Dickson, *The Life And Inventions of Thomas Alva Edison* (Boston: Thomas Crowell, 1894), 37–39; Ray Phillips, *Edison's Kinetoscope and its Films: A History to 1896* (Westport, CT: Greenwood Press, 1997), 43.

48. Gordon Hendricks, *The Kinetoscope* (New York: Theodore Gaus' Sons, 1966), 60–66; *Edison Phonographic News*, cited in Ray Phillips, *Edison's Kinetoscope and Its Films: A History to 1896* (New York: Praeger, 1997), 28.

49. Hendricks, *The Kinetoscope*, 7, 94.

50. Phillips, *Edison's Kinetoscope*, 31, 33, 61–2; Hendricks, *The Kinetoscope*, 55, 128–29; Paul Spehr, "William Dickson," in Able, *Encyclopedia of Early Cinema*, 186.

51. Mannoni, *The Great Art of Light and Shadow*, 427–33; Theresa Collins, *Thomas Edison and Modern America* (New York: St. Martin's, 2002), 23–24; Gordon Hendricks, *Beginnings of the Biograph* (New York: Beginnings of the American Film, 1964), 9–15, 30–33.

52. Film competitors included Edison and Biograph, as well as American Vitagraph, Lubin, Selig, and several European companies, including Pathé. Richard Abel, *Americanizing the Movies and "Movie-Mad" Audiences, 1910–1914* (Berkeley: University of California Press, 2002), 134; Raymond Fielding, "Hale's Tours:

Ultrarealism in the Pre-1910 Motion Picture," *Cinema Journal* 10, no. 1 (Autumn 1970): 34–47.

53. *The Edison Kinetoscope Price List*, August 1895 (Cincinnati: Ohio Phonograph Co, 1895), 7; Phillips, *Edison's Kinetoscope*, 69, 140–41, 144; "Edison Films" FTA, DVDs 27, 30, 618; "1890s Films," FTA, VCR 13349.

54. Musser, *The Emergence of the Cinema*, 17, argues for continuity with the past in terms of the aesthetic appeal of early cinema while many historians focus on invention. John Fell, *Film and Narrative Tradition* (Norman: University of Oklahoma Press, 1974) sees that early cinema borrowed from comic strips, dime novels, songs, magic lanterns, and theater. Robert Allen finds a connection with vaudeville in his *Vaudeville and Film, 1895–1915* (New York: Arno, 1980).

55. Abel, *Americanizing the Movies and "Movie-Mad" Audiences*, 134; Charlotte Herzog, "The Movie Palace and the Theatrical Sources of Its Architectural Style," *Cinema Journal* 20, no. 2 (Spring 1981): 24–28; Barton Currie, "The Nickel Madness," *Harper's Weekly*, August 24, 1907, 1246.

56. Jean-Gabriel Tarde, *Laws of Imitation* (New York, 1903), 239, 322–44; Tarde, *La Psychologie économique*, vol. 2 (Paris, 1902), 151–56, 256, 264.

57. Tom Gunning, *D. W. Griffith and the Origins of American Narrative Film* (Urbana: University of Illinois Press, 1991), 56–65, 85–86; Tom Gunning, "The World as Object Lesson: Cinema Audiences, Visual Culture and the St. Louis World's Fair, 1904," *Film History* 6, no. 4, (Winter 1994): 423.

58. Tom Gunning, "The Cinema of Attraction(s): Early Film, Its Spectator and the Avant-Garde," in *Early Cinema: Space, Frame, Narrative*, ed. Thomas Elsaesser (London: British Film Institute, 1990), 56–62; Tom Gunning, "Rethinking Early Cinema: Cinema of Attraction and Narrativity," in *Cinema of Attractions Reloaded*, ed. Wanda Strauven, 389–416 (Amsterdam: Amsterdam University Press, 2006); Charles Musser, "A Cinema of Contemplation, Cinema of Discernment," in Strauven, *Cinema of Attractions Reloaded*, 176.

59. Musser, *The Emergence of Cinema*, 128; Frank Kessler, "Trick Films," in Able, *Encyclopedia of Early Cinema*, 643–45; Samantha Barbas, *Movie Crazy: Fans, Stars, and the Cult of Celebrity* (New York: Palgrave, 2001), 11–14; Kemp Niver, *The First Twenty Years: A Segment of Film History* (Los Angeles: Locare Research Group, 1968), 19–34 38–39; *Early Trick Films*, FTA, VA 1008; *Porter's Films, 1898–03*, FTA, VA 1528 (for a series of trick films); *Wonderful Wizard of Oz, 1910*, FTA, DVD 618; *Winsor McCay Cartoons*, FTA, DVD 5951.

60. Musser, *The Emergence of Cinema*, 225–63; Collins, *Thomas Edison and Modern America*, 22; Andrea Stulman Dennett and Nina Warnke, "Disaster Spectacles at the Turn of the Century," *Film History* 4, no. 2 (1990): 101–11; Niver, *The First Twenty Years*, 42; "Actuality Films," FTA, DVD 30; DVD 2851; *Life of an American Policeman*, FTA, DVD 2851; *Life of an American Fireman* (Edison, 1903) and *The Great Train Robbery* (Edison, 1903), FTA, VA 1528; *The Suburbanite* (Biograph, 1904), FTA, DVD 618.

61. Henry Jenkins, *The Wow Climax: Tracing the Emotional Impact of Popular Culture* (New York: New York University, 2007), 4–10.

62. Musser, *The Emergence of Cinema*, chap. 10, offers a good summary.

63. Abel, *Americanizing the Movies and "Movie-Mad" Audiences*, 17–19; Douglas Gomery, *Shared Pleasures: A History of Movie Presentation in the United States* (Madison: University of Wisconsin Press, 1992); Tom Gunning, "Motion Picture Patents Company," in Able, *Encyclopedia of Early Cinema*, 447–48; Tom Gunning, "Cinema of Attractions," in Able, *Encyclopedia of Early Cinema*, 124–26.

64. Abel, *Americanizing the Movies and "Movie-Mad" Audiences*, chaps. 2, 3, and 4; Richard Abel, *The Red Rooster Scare: Making Cinema American, 1900–1910* (Berkeley: University of California Press, 1999), chaps. 3, 5; Ben Singer, *Melodrama and Modernity: Early Sensational Cinema and its Contexts* (New York: Columbia University Press, 2001), 149–88; Bill Brown, ed. *Reading the West: An Anthology of Dime Westerns* (Boston: Bedford, 1997).

65. An early example of a movie serial is *Hazards of Helen* (Kalem, 1915), FTA, DVD 618; see also David Zinman, *Saturday Afternoon at the Bijou* (New Rochelle, NY: Arlington House, 1973), 288–300; Jim Harmon and Donald Glut, *Great Movie Serials: Their Sound and Fury* (Garden City, NJ: Doubleday, 1972), 2–5; Kalton Lahue, *Continued Next Week: A History of the Moving Picture Serial* (Norman, OK: University of Oklahoma Press, 1964), chaps. 4–10.

66. Gunning, *D. W. Griffith and the Origins of American Narrative Film*, 85, 89–91.

67. Abel, *Americanizing the Movies and "Movie-Mad" Audiences*, 232–33; Samantha Barbas, *Movie Crazy: Fans, Stars, and the Cult of Celebrity* (New York: Palgrave, 2001), 15–28, 35–57; Richard DeCordova, *Picture Personalities: The Emergence of the Star System in America* (Urbana: University of Illinois Press, 1990), 85–90.

68. Herzog, "Movie Palace,"15–21; Abel, *Americanizing the Movies and "Movie-Mad" Audiences*, 51. Douglas Gomery, *Shared Pleasures: A History of Movie Presentation in the United States* (Madison: University of Wisconsin Press, 1992), 34–40, 69–87; Steven Ross, *Working-Class Hollywood: Silent Film and the Shaping of Class in America* (Princeton, NJ: Princeton University Press, 1998).

69. Walter Benjamin, *The Work of Art in the Age of its Technological Reproducibility, and Other Writings on Media*, ed. Michael Jennings, Brigid Doherty, and Thomas Levin (Cambridge, MA: Harvard University Press, 2008), 22, 26, 29, 35, 37.

70. Hugo Münsterberg, *The Photoplay* (New York: Appleton, 1916), 157, cited in Kay Sloan, "The Loud Silents: Origins of the Social Problem Film," in *Movies and American Society*, ed. Steven Ross (New York: Blackwell, 2002), 52–53.

71. Benjamin, *The Work of Art*, 39–41; Jütte, *A History of the Senses*, 300–1.

72. Benjamin, *The Work of Art*, 22–5, 27.

73. Guy Debord, *Society of the Spectacle* (New York: Zone Books, 1995), 12–13, 18.

## Chapter 7

1. David Sloan Wilson, *Darwin's Cathedral: Evolution, Religion and the Nature of Society* (Chicago: University of Chicago Press, 2002).

2. Peter Laslett, *The World We Have Lost* (New York: Scribner, 1984), chap. 2; Joffre Dumazedier, *Sociology of Leisure* (New York: Elsevier Scientific, 1974), 34; Christina Hole, *British Folk Customs* (London: Hutchinson, 1976) 63, 137.

3. Michael Judge, *The Dance of Time: The Origins of the Calendar* (New York: Arcade Publishing, 2004) 6, 56, 143, 145–59, 199–203; Hole, *British Folk Customs*, 137.

4. Charles Caraccioli, *An Historical Account of Stourbridge* (Cambridge, 1773), 20–21, cited in Robert Malcolmson, *Popular Recreations in English Society* (Cambridge: Cambridge University Press, 1973), 21; William Addison, *English Fairs and Markets* (London: Batsford, 1953), 95–225.

5. Mark Judd, "Popular Culture and the London Fairs, 1800–1860," in *Leisure in Britain*, ed. John Walton and James Walvin (Manchester, UK: Manchester University Press, 1983), 2–25; Jack Santino, *All Around the Year: Holidays and Celebrations in American Life* (Urbana: University of Illinois Press, 1994), 90, 145–64.

6. Peter Burke, *Popular Culture in Early Modern Europe* (New York: Harper, 1978), 178–204; Emmanuel Le Roy Ladurie, *Carnival in Romans* (New York: G. Braziller, 1979), 305–24.

7. Burke, *Popular Culture*, 178–204; Mikhail Bakhtin, *Rabelais and His World* (Bloomington: Indiana University Press, 1984), chap. 1. For American variations on Carnival, see Santino, *All Around the Year*, 88–96.

8. Marie Luise Gothein, *History of Garden Art: From the Earliest Times to the Present Day*, vol. 1 (London: J. W. Dent, 1928), 25–30; Julia Berrall, *The Garden: An Illustrated History from Ancient Egypt to the Present Day* (London: Thames & Hudson, 1966), 35–39.

9. Robert Berger, *In the Garden of the Sun King* (Washington DC: Dumbarton Oaks, 1985); Karen Jones and John Wills, *The Invention of the Park* (Cambridge: Polity, 2005) 9–25.

10. Kenneth Woodbridge, *The Stourhead Landscape* (London: The National Trust, 2002); Tom Williamson, *Polite Landscapes* (Baltimore, MD: The Johns Hopkins University Press, 1995); John D. Hunt, *Gardens and the Picturesque* (Cambridge, MA: MIT Press, 1992); Jones and Wills, *The Invention of the Park*, chap. 2; Terrence Young, "Grounding the Myth," in *Theme Park Landscapes: Antecedents and Variations*, ed. Terence Young and Robert Riley (Washington, DC: Dumbarton Oaks, 2002), 6–7.

11. Edward Harwood, "Rhetoric, Authenticity, and Reception: The Eighteenth-Century Landscape Garden, the Modern Theme Park, and their Audiences," in Young and Riley, *Theme Park Landscapes*, 66.

12. Susan Stewart, *On Longing: Narratives of the Miniature, the Gigantic, the Souvenir, the Collection* (Durham, NC: Duke University Press, 1993), 75.

13. Jones and Will, *The Invention of the Park*, 38–39; Neville Braybrooke, *London Green; The Story of Kensington Gardens, Hyde Park, Green Park & St. James's Park* (London: Gollancz, 1959), 26–27; Judith Adams, *The American Amusement Park Industry: A History of Technology and Thrills* (Boston: Twayne, 1991), 3–8.

14. Roy Porter, *English Society in the Eighteenth Century* (New York: Penguin, 1982), 242–50; T. Lea Southgate "Music at the Public Pleasure Gardens of the Eighteenth Century," *Proceedings of the Musical Association*, 38th Sess. (1911–1912), 141–42; Warwick Wroth, *The London Pleasure Gardens of the Eighteenth Century* (London: Macmillan, 1896).

15. James Granville Southworth, *Vauxhall Gardens* (New York: Columbia University Press, 1941), 36–71; David Coke, *Vauxhall Gardens* (New Haven, CT: Yale

University Press, 2011); John Quinlan, "Music as Entertainment in 18th-Century London," *Musical Times*, July 1, 1932, 612–14.

16. Heath Schenker, "Pleasure Gardens, Theme Parks, and the Picturesque," in Young and Riley, *Theme Park Landscapes*, 69–89; Roy Rosenzweig and Elizabeth Blackmar, *The Park and the People: A History of Central Park* (Ithaca, NY: Cornell University Press, 1992); T. M. Garrett, "A History of Pleasure Gardens in New York City, 1700–1865" (PhD diss., New York University, 1978).

17. R. S. Neale, *Bath, 1680–1850* (London: Routledge, 1981); Peter Borsay, *The Image of Georgian Bath, 1700–2000* (Oxford: Oxford University Press, 2000); James Stevens Curl, "Spas and Pleasure Gardens of London, from the Seventeenth to the Nineteenth Centuries," *Garden History* 7, no. 2 (Summer 1979): 27–68; Warwick Wroth, *The London Pleasure Gardens of the Eighteenth Century* (1896; London: Macmillan, 1979).

18. Phyllis Hembry, *The English Spa* (London: Athlone Press, 1990), 1–25, 99, 135–37, 302–5.

19. Curl, "Spas and Pleasure Gardens of London," 47, 55; Kristina Taylor, "The Oldest Surviving Pleasure Garden in Britain: Cold Bath, Near Tunbridge Wells in Kent," *Garden History* 28, no. 2 (Winter 2000): 277–82.

20. Anthony Hern, *The Seaside Holiday: The History of the English Seaside Resort* (London: Cresset Press, 1967), 7–11, 45–55, 140–45; Sue Berry, "Pleasure Gardens in Georgian and Regency Seaside Resorts: Brighton, 1750–1840," *Garden History* 28, no. 2 (Winter 2000): 222–30; John K. Walton, *The English Seaside Resort: A Social History 1750–1914* (Leicester, UK: Leicester University Press, 1983), 158–61; Alain Corbin, *The Lure of the Sea* (Cambridge: Polity, 1994); Sue Farrant, *Georgian Brighton, 1750–1820* (Brighton, UK: University of Sussex, 1980), chap. 1.

21. George Waller, *Saratoga: Saga of an Impious Era* (Englewood Cliffs, NJ: Prentice Hall, 1966), 56–108; Jon Sterngass, *First Resorts: Pursuing Pleasure at Saratoga Springs, Newport, and Coney Island* (Baltimore, MD: The Johns Hopkins University Press, 2001), 241; Theodore Corbett, *The Making of American Resorts: Saratoga Springs, Ballston Spa, Lake George* (New Brunswick, NJ: Rutgers University Press, 2001), 79; Cindy S. Aron, *Working at Play: A History of Vacations in the United States* (New York: Oxford University Press, 1999).

22. John Sears, *Sacred Places: American Tourist Attractions in the Nineteenth Century* (New York: Oxford University Press, 1989), 28, 185–88; Sterngass, *First Resorts*, 7–74, 1, 117–45, 204–20, 227.

23. Patrick Beaver, *The Crystal Palace: A Portrait of Victorian Enterprise* (Chilchester, UK: Phillimore, 1986), 47, 57; David Nasaw, *Going Out: The Rise and Fall of Public Amusements* (New York: Basic Books, 1993), chaps. 2 and 3; Robert Bogdan, *Freak Show* (Chicago: University of Chicago Press, 1988), 50–51; Robert Rydell, *World of Fairs: The Century-of-Progress Expositions* (Chicago: University of Chicago Press, 1993), chaps. 1, 4, and 5.

24. Robert Rydell, John Findling, and Kimberly Pell, *Fair America* (Washington DC: Smithsonian Institution Press, 2000), 37–39; "The Great Wheel at Chicago," *Scientific American*, July 1, 1893, 234; Adams, *The American Amusement Park Industry*, 31–35; James Gilbert, *Perfect Cities: Chicago's Utopias of 1893* (Chicago: University of Chicago Press, 1991), 108–18; Ed MCullough, *World's Fair Midways* (New York: Exposition Press, 1966), 24–71.

25. Tom Gunning, "From the World as Object Lesson: Cinema Audiences, Visual Culture and the St. Louis World's Fair, 1904," *Film History*, 6, 4, (Winter, 1994): 425–35.

26. Winterthur Museum, Collection 45, file 2; Gunning, "St. Louis World's Fair," 435–37.

27. Gunning, "St. Louis World's Fair," 437; Rydell, *All the World's a Fair*, 159.

28. Adams, *The American Amusement Park Industry*, 7; John K Walton, *Blackpool* (Edinburgh: Edinburgh University Press, 1998), 14–22.

29. Buffalo Bill's show appeared at many venues in the United States and abroad. It was a veritable variety show of western-themed attractions, including a quarter-mile race of an Indian, Mexican, and cowboy; a reenactment of the robbery of the Deadwood stagecoach; exhibitions of sharpshooters, and the Pawnee's Indian scalp and war dance. William Deahl, "Buffalo Bill's Wild West Show in New Orleans," *Louisiana History: The Journal of the Louisiana Historical Association* 16, no. 3 (Summer 1975): 289–98.

30. Gary Cross and John Walton, *The Playful Crowd: Pleasure Places in the Twentieth Century* (New York: Columbia University Press, 2005), chaps. 1–4. The bizarre Elephant Hotel (standing between 1884 and 1896) is described in Clay Lancaster, *Architectural Follies in America* (Rutland, VT: Charles Tuttle, 1960), 194–96.

31. Gary Kyriazi, *The Great American Amusement Parks* (Secaucus, NJ: Citadel Press, 1976), 34–42.

32. Ibid., 47–57; "New Steeplechase to cost $1,000,000," *New York Times*, December 29, 1907; "Summer Amusement Parks," *New York Times*, August 16,

1908; Tilyou quotation from Reginald Kauffman, "Why is Coney?" *Hampton's Magazine*, August 1909, 224.

33. David Nye, *American Technological Sublime* (Cambridge, MA: MIT Press, 1996).

34. Oliver Pilat and Jo Ranson, *Sodom by the Sea, An Affectionate History of Coney Island* (Garden City, NY: Doubleday, 1941), 144–46; Woody Register, *Kid of Coney Island: Fred Thompson and the Rise of American Amusements* (New York: Oxford University Press, 2001), 92, 132–33.

35. Adams, *The American Amusement Park Industry*, 48. Register; *Kid of Coney Island*, 121.

36. Stephen Weinstein, "The Nickel Empire: Coney Island and the Creation of Urban Seaside Resorts in the United States" (PhD diss., Columbia University, 1984), 249.

37. Barr Ferree, "The New Popular Resort Architecture, Dreamland, Coney Island," *Architects' and Builders' Magazine*, August 1904, 499; Weinstein, "The Nickel Empire," 220–23. See also David R. Francis, *The Universal Exposition of 1904* (St. Louis, MO: St. Louis Purchase Exposition Co., 1913), 567, for a description of the "Creation."

38. Jeffrey Stanton, "Coney Island—Dreamland," April 1998, 1–11, www .westland.net/coneyisland/articles/dreamland.htm.

39. Frederic Thompson, "Amusing the Million," *Everybody's Magazine*, September 1908, 378–86; Thompson, "Amusement Architecture," *Architectural Review* 16, no. 7 (July 1909): 87–89; Michele Bogart, *Public Sculpture and the Civic Ideal in New York City, 1890–1930* (Chicago: University of Chicago Press, 1989), 248–57.

40. Richard Le Gallienne, "The Human Need for Coney Island," *Cosmopolitan*, July 1905, 243; Cross and Walton, *The Playful Crowd*, chap. 3.

41. John Kasson, *Amusing the Million* (New York: Hill and Wang, 1978), 70; William Mangels, *The Outdoor Amusement Industry from Earliest Times to the Present* (New York: Vantage, 1952), 165; Daniel Boorstin, *The Image: A Guide to Pseudo-Events in America*(New York: Harper & Row, 1961).

42. Kyriazi, *The Great American Amusement Parks*, 67–70; Stanton, "Coney Island—Dreamland,"1–11.

43. Julian Hawthorne, "Some Novelties at Buffalo Fair," *Cosmopolitan*, September 1901, 490–91, and "Pan-American Exposition," promotional brochure (Buffalo, NY, 1901), 29; Register, *Kid of Coney Island*, 71, 74–76; Albert Paine,

"The New Coney Island," *Century*, August 1904, 544; *Official Catalogue and Guide Book to the Pan-American Exposition, Buffalo, N.Y.* (New York: Charles Ehrhart, 1901), 44.

44. Rollin Lynde Hartt, "The Amusement Park," *Atlantic Monthly*, May 1907, 675–76, republished in Rollin Hartt, *People at Play* (New York: Houghton Mifflin, 1909), 75–76; "Great New Dreamland at Coney Island this Year," *New York Times*, April 23, 1905.

45. "Luna Park Opens," *New York Times*, May 7, 1905; "Roof Gardens and Summer Theatrical Offerings," *New York Times*, June 11, 1905; Michael Immerso, *Coney Island: The People's Playground* (Piscataway, NJ: Rutgers University Press, 2002), 71; "Luna Park," *New York World*, July 20, 1902; Francis, *The Universal Exposition of 1904*, 600, for description of the Galveston Flood show.

46. Lynn Sally, *Fighting the Flames: Spectacle Performance of Fire at Coney Island* (New York: Routledge, 2006), 52–63; Rem Koolhass, *Delirious New York* (New York: Oxford University Press, 1978), 49; Stanton, "Coney Island — Dreamland."

47. Andrea Stulman Dennett and Nina Warnke, "Disaster Spectacles at the Turn of the Century," *Film History* 4, no. 2 (1990): 101–11; "Special Features of White City," *White City Magazine*, May 1905, 52; Sally, *Fighting the Flames*, 89–97.

48. Marvin Zuckerman, *Sensation and Risky Behavior* (Washington DC: American Psychological Association, 2007); Marvin Zuckerman, *Behavioral Expressions and Biosocial Bases of Sensation Seeking* (New York: Cambridge University Press, 1993); Brian Sutton-Smith, *Ambiguity of Play* (Cambridge, MA: Harvard University Press, 2001); Ivana Hromatko and Ana Butkovic, "Sensation Seeking and Spatial Ability in Athletes: An Evolutionary Account," *Journal of Human Kinetics* 21 (2009): 5–13.

49. Garrett Soden, *Falling: How Our Greatest Fear Became our Greatest Thrill — A History* (New York: Norton, 2003), 1–20, 79, 108 (quotation on 38).

50. Kasson, *Amusing the Million*, 59, 60; Adams, *The American Amusement Park Industry*, 45.

51. Hartt, "The Amusement Park," 675.

52. Mangels, *The Outdoor Amusement Industry*, 37–50, 137, 163; Soden, *Falling*, 22–29, 160–62.

53. Todd Throgmorton, *Roller Coasters of America* (Oscelola, WI: Motor Books, 1994), 26–27; Todd Throgmorton, *Roller Coasters: United States and Canada* (Jefferson, NC: McFarlane, 2000), 1–18. See also the following articles

from *Scientific American*: "Looping the Double Loop," 90, July 8, 1905, 493; "Leap-Frog Railway," 93, July 8, 1905, 29–30; "Mechanical Joys of Coney Island," 99, August 15, 1908, 101; "Mechanical Side of Coney Island," 103, August 6, 1911, 104–13; "How High Can You Go," *New Yorker*, August 30, 2004, 48.

54. Immerso, *Coney Island*, 137–47; Pilat and Ranson, *Sodom by the Sea*, 220–25; Throgmorton, *Roller Coasters*, 13–16.

55. Michael DeAngelis, "Orchestrated (Dis)orientation: Roller Coasters, Theme Parks, and Postmodernism," *Cultural Critique* 37 (Autumn 1997): 113; Jean Baudrillard, "Consumer Society," in his *Selected Writings* (Stanford, CA: Stanford University Press, 1988), 49.

56. Anson Rabinbach, *The Human Motor: Energy, Fatigue, and the Origins of Modernity* (New York: Basic, 1990) explores industrial fatigue in this period.

57. "Mechanical Joys of Coney Island," *Scientific American*, 109; Lauren Rabinovitz, "Urban Wonderlands: Siting Modernity in Turn of the Century Amusement Parks," *European Contributions to American Studies* 45, no. 1 (2001): 88; Koolhass, *Delirious New York*, 27, 42; Kasson, *Amusing the Million*, 72–74.

58. Nye, *American Technological Sublime*.

59. Henry Pain ad, *Harper's Bazaar*, June 23, 1894, 27, 25.

60. David Francis, *Cedar Point* (Charleston, SC: Arcadia, 2004), 20–26.

61. "How Ponce de Leon was Made into an Amusement Park," *Atlanta Constitution*, July 17, 1904.

62. "Bumps," *White City Magazine*, February 1905, 28; "Johnstown Flood at White City," *White City Magazine*, March 1905, 13; "Shooting the Chutes," *White City Magazine*, May 1905, 30; Scott Newman, "Boundless Pleasures: Young Chicagoans, Commercial Amusements, and the Revitalization of Urban Life, 1900–1930" (PhD diss., Loyola University Chicago, 2004), 24–38; Stan Bark, "Paradises Lost," *Chicago History* 22 (March 1993): 26–49.

63. John Walton, "Social Development of Blackpool, 1788–1912" (PhD diss., Lancaster University, 1974), 327–29; Peter Bennett, *Century of Fun* (Blackpool: Blackpool Pleasure Beach, 1996), 12–25; Cross and Walton, *The Playful Crowd*, chap.1; John Walton, *Riding on Rainbows: Blackpool Pleasure Beach and Its Place in British Popular Culture* (St. Albans: Skelter Publishing, 2007).

64. Register, *Kid of Coney Island*, 141; Jones and Wills, *Invention of the Park*, 99–100; Cross and Walton, *The Playful Crowd*, chaps. 4 and 5.

65. Photo collection of the Brooklyn Historical Society, V1974.22.6.40; V1974.19.1.4; E. V. Lucas, *Roving East and Roving West* (London: Methuen, 1921),

111. On the adult evocation of "wondrous innocence" in children, see Gary Cross, *The Cute and the Cool: Wondrous Innocence and Modern American Children's Culture* (New York: Oxford University Press, 2003).

66. Hartt, "The Amusement Park," 675.

67. Thompson, "Amusing the Million," 385–87; Frederic Thompson, "Amusing People," *Metropolitan Magazine*, July 106, 602–3; Frederic Thompson, "The Summer Show," *Independent*, July 6, 1907, 1461.

68. Edward Tilyou, "Human Nature with the Brakes Off—Or: Why the Schoolma'am Walked into the Sea," *American Magazine*, July 1922, 19, 92; Register, *Kid of Coney Island*, 12, 16.

69. Darrin McMahon, *Happiness: A History* (New York: Atlantic Monthly Press, 2006), 12–13; Peter N. Stearns, *Satisfaction Not Guaranteed: Dilemmas of Progress in Modern Society* (New York: New York University Press, 2012), chap. 2.

## Chapter 8

1. Gary Cross, *All-Consuming Century: Why Commercialism Won in Modern America* (New York: Columbia University Press, 2000).

2. Stephen Kline, Nick Dyer-Withford, Greig De Peuter, *Digital Play: The Interaction of Technology, Culture, and Marketing* (Montreal: McGill-Queen's University Press, 2003), 84–108.

3. Hervé Fischer, *Digital Shock: Confronting the New Reality* (Montreal: McGill-Queen's University Press, 2006).

4. Jasper Woodroof and G. Frank Phillips, *Beverages: Carbonated and Noncarbonated* (Westport, CT: Avi Publishing, 1974), 5–6; Anne Cooper Funderburg, *Sundae Best: A History of Soda Fountains* (Bowling Green, OH: Bowling Green State University Press, 2002), 123–53; Beverage World, *Coke's First 100 Years* (Great Neck, NY: Beverage World, 1986), 105–51, 164; "History of Packaging and Canning," Answers.com, http://www.answers.com/topic/history-of-packaging-and-canning.

5. Kelly Brownell and Katherine Horgen, *Food Fight: The Inside Story of the Food Industry, America's Obesity Crisis and What We Can Do About It* (New York: McGraw-Hill, 2004), 28–30.

6. Harvey Levenstein, *Paradox of Plenty: A Social History of Eating in Modern America* (New York: Oxford University Press, 1993), 106–7; Robert Phipps, *The Swanson Story* (Omaha, NE: Swanson Foundation, 1977), 77–80; Karol Ann Mar-

ling, *As Seen on TV: The Visual Culture of Everyday Life in the 1950s* (Cambridge, MA: Harvard University Press, 1994), 232–34.

7. William Walsh, *The Rise and Decline of the Great Atlantic and Pacific Tea Company* (Secaucus, NJ: Stuart, 1986), 34–35; Richard Longstreth, *The Drive-In, the Supermarket, and the Transformation of Commercial Space in Los Angeles, 1914–1941* (Cambridge, MA: MIT Press, 1999), 82–111; Andrew Smith, *Eating History* (New York: Columbia University Press, 2009), 177–79.

8. Smith, *Eating History*, 128–30; Tim Richardson, *Sweets* (London: Bloomsbury, 2002), 23–25.

9. Richard Pillsbury, *From Boarding House to Bistro: The American Restaurant Then and Now* (Boston: Unwin, 1990), 48–105; Andrew Smith, *Hamburger: A Global History* (London: Reaktion Books, 2008), 15–20; David Hogan, *Selling 'Em by the Sack: White Castle and the Creation of American Food* (New York: NYU Press, 1997), chap. 1; John Love, *McDonald's: Behind the Arches* (New York: Bantam, 1995), 25, 160; Smith, *Eating History*, 219–28.

10. One of the best recent analyses of this is David Kessler, *End of Overeating* (New York: Rodale, 2009), 67–134.

11. Sungook Hong, *Wireless: From Marconi's Black-Box to the Audion* (Cambridge, MA: MIT Press, 2001); Steven Wurtzler, *Electric Sounds: Technological Change and the Rise of Corporate Mass Media* (New York: Columbia University Press, 2007); Christopher Sterling, *The Rise of American Radio*, vol. 1 (New York: Routledge, 2007); Susan Douglas, *Inventing American Broadcasting, 1899–1922* (Baltimore, MD: Johns Hopkins University Press, 1987).

12. Oliver Read and Walter Welch, *From Tin Foil to Stereo: Evolution of the Phonograph* (New York: Boobs-Merrill, 1976), 224, 256; A. J. Millard, *America on Record* (New York: Cambridge University Press, 1995), 136–57; Roland Gelatt, *The Fabulous Phonograph 1877–1977* (New York: Macmillan, 1977), 218; Michael Chanan, *Repeated Takes: A Short History of Recording and its Effects on Music* (London, Verso, 1995), 38–39.

13. Gelatt, *The Fabulous Phonograph*, 220–44; Chanan, *Repeated Takes*, 56–67; Millard, *America on Record*, 145.

14. Gelatt, *The Fabulous Phonograph*, 245–67; Mark Coleman, *Playback: From the Victrola to MP3, 100 Years of Music, Machines, and Money* (New York: Da Capo, 2003), 43.

15. Timothy Day, *A Century of Recorded Music* (New Haven, CT: Yale University Press, 2000), 19; Coleman, *Playback*, 39, 59–68, 76–85.

16. Coleman, *Playback*, 57–59; Greg Milner, *Perfecting Sound Forever: An Aural History of Recorded Music* (London: Faber and Faber, 2009), 109–12; Simon Reynolds, *Retromania: Pop Culture's Addiction to its Own Past* (New York: Faber and Faber, 2012), 60–75.

17. Coleman, *Playback*, 98–101; Day, *A Century of Recorded Music*, 20; Tom Anderson, *Making Easy Listening: Material Culture and Postwar American Recording* (Minneapolis: University of Minnesota Press, 2006), 149, 170–78.

18. Coleman, *Playback*, xv (quotation), xvi, 159–63; Day, *A Century of Recorded Music*, 2–20; Milner, *Perfecting Sound Forever*, 191–94.

19. Anderson, *Making Easy Listening*, 7–12, 24, 34–37, 44, 111; Jacques Attali, *Noise: The Political Economy of Music* (Minneapolis: University of Minnesota Press, 1985), 87, 128; David Suisman, *Selling Sound: The Commercial Revolution in American Music* (Cambridge, MA: Harvard University Press), 282.

20. See Jan Jarvis, "Notes on Muzak," in *The Phonograph and Our Musical Life*, ed. H. Wiley Hitchcock (Albany: SUNY Press 1977), 13–15; Attali, *Noise*, 111–12; Suisman, *Selling Sound*, 256–58.

21. "Charge-Coupled Device," Wikipedia, http://en.wikipedia.org/wiki/Charge-coupled_device#History; Martin Lister, ed., *The Photographic Image in Digital Culture* (London: Routledge, 1995).

22. Millard, *America on Record*, 149.

23. Eric Lichtenfeld, *Action Speaks Louder: Violence, Spectacle, and the American Action Movie* (Westport, CT: Praeger, 2004), 17.

24. Ibid., 22–25; John Taylor, "Dirty Harry," in *Movies of the Seventies*, ed. Ann Lloyd (London: Orbis, 1984), 172–73.

25. Derek Elley, "Martial Arts Films," in Lloyd, *Movies of the Seventies*, 190–91; Yvonne Tasker, *Spectacular Bodies: Gender, Genre, and the Action Cinema* (New York: Routledge, 1993), 2–3, 79–80.

26. Lichtenfeld, *Action Speaks Louder*, 186–87; Larent Bouzereau, *Ultra-Violent Movies* (Secaucus: Carol Publishing, 1996), 91; Thomas Leitch, "Aristotle vs. the Action Film," in *New Hollywood Violence*, ed. Steven Schneider (Manchester, UK: Manchester University Press, 2004), 116–17.

27. Barry Atkins, *More Than a Game: The Computer Game as Fictional Form* (Manchester, UK: Manchester University Press: 2003), chap. 1; Kline et al., *Digital Play*, 84–108.

28. David Sheff, *Game Over: How Nintendo Conquered the World* (Wilton, CT: Gamepress, 1999), 150–57; J. C. Herz, *Joystick Nation* (Boston: Little Brown,

1997), 14–22, 33–37, 55; Leonard Herman, *Phoenix: The Fall and Rise of Video Games* (Springfield, NJ: Rolenta Press, 2001), 89–99; Steven Malliet and Gust de Meyer, "The History of the Video Game," in *Handbook of Computer Games Studies*, ed. Joost Raessens and Jeffrey Goldstein (Cambridge, MA: MIT Press, 2005), 26–28.

29. Kline et al., *Digital Play*, 128–150.

30. According to the Entertainment Software Association (ESA), representing the video game industry, in 1997 more than half of all video game players were younger than 18. By 2005 the average age had risen to 33 and by 2011 to 37, roughly charting the aging of the Gen Xers who were introduced to video games as children. Entertainment Software Association, http://www.theesa.com/archives/files /Essential%20Facts%202006.pdf, and "Essential Facts about the Computer and Video Game Industry: 2011 Sales, Demographic and Usage Data," http://www .theesa.com/facts/pdfs/ESA_EF_2011.pdf.

31. Jaron Lanier, *You are Not a Gadget* (New York: Knopf, 2010); Jim Blascovich and Jeremy Bailenson, *Infinite Reality: Avatars, Eternal Life, New Worlds, and the Dawn of the Virtual Revolution* (New York: HarperCollins, 2011), 170–81; E. Aboujaoude, L. M. Koran, N. Gamel, et al. "Potential Markers for Problematic Internet Use: A Telephone Survey of 2,513 Adults," *CNS Spectrums* 11 (2006): 924–30; Robert E. Kraut et al., "Social Impact of the Internet: What Does it Mean?" *Communications of the ACM*, 41 (1998): 21–22.

32. Gary Cross, *Men to Boys: The Making of Modern Immaturity* (New York: Columbia University Press, 2008), 212–25; "Lost in an Online Fantasy World As Virtual Universes Grow, So Do Ranks of the Game-Obsessed," *Washington Post*, August 18, 2006.

33. James Gleick, *Faster: The Acceleration of Just about Everything* (New York: Pantheon, 1999), 177, 185–90.

34. J. MacLachlan and M. Logan, "Camera Shot Length in TV Commercials and Their Memorability and Persuasiveness," *Journal of Advertising Research* 33 (1993): 57–63.

35. Simon Gottschalk, "Speed Culture: Fast Strategies in Televised Commercial Ads," *Qualitative Sociology* 22 (1999): 312; Wendy Parkins and Geoffrey Craig, *Slow Living* (Oxford: Berg, 2006), 38.

36. Todd Throgmorton, *Roller Coasters of America* (Osceola, WI: Motorbooks, 1994), 13–16.

37. Gary Cross and John Walton, *The Playful Crowd: Pleasure Places in the*

*Twentieth Century* (New York: Columbia University Press, 2005), chap. 5. Note also Brenda Brown, "Landscapes of Theme Park Rides: Media, Modes, Messages," in *Theme Park Landscapes: Antecedents and Variations*, ed. Terence Young and Robert Riley (Washington, DC: Dumbarton Oaks, 2002), 358–62.

38. Todd Throgmorton, *Roller Coasters* (Jefferson, NC: McFarland, 1993), 32–33; Scott Rutherford, *The American Roller Coaster* (Osceola, WI: MBI Publishing, 2000), 102–3.

39. Throgmorton, *Roller Coasters*, 26; Robert Cartmell, *The Incredible Scream Machine* (Bowling Green, OH: Bowling Green State University Popular Press, 1987), 182; "Matterhorn," *News from Disneyland*, December 12, 1959.

40. Roller Coaster Data Base, http://www.rcdb.com/glossary.htm; Rutherford, *The American Roller Coaster*, 82.

41. http://channel.nationalgeographic.com/channel/supercoasters/facts.html; Throgmorton, *Roller Coasters*, 35; Robert Coker, *Roller Coasters* (New York: MetroBooks, 2002), 8–11.

42. Details are in the Cedar Point web page, http://www.cedarpoint.com /rides/Roller-Coasters/Top-Thrill-Dragster; "Roller Coasters: A Steep Upswing," *Business Week*, June 21, 1999, 8; "The Year of the Roller Coaster," *Lighting Dimensions*, November 1, 2003, 11; "The Thrill isn't Gone," *The New Yorker*, August 30, 2004, 48.

43. "Body Wars," *Disney News*, Spring 1989, 36; "Rock 'n Roller Coaster," *Eyes and Ears* (Walt Disney World in-house newsletter), April 23, 1998, 1–3; "On Track," *Disney Magazine*, Fall 1998, 44–47.

**Chapter 9**

1. Gene Wallenstein, *The Pleasure Instinct: Why We Crave Adventure, Chocolate, Pheromones, and Music* (New York: Doubleday, 2009), 190–204.

2. David Rosner and Gerald Markowitz, "A 'Gift of God'? The Public Health Controversy over Leaded Gasoline During the 1920s," *American Journal of Public Health* 75 (1985): 344–52.

3. David Kessler, *The End of Overeating: Taking Control of the Insatiable American Appetite* (New York: Rodale, 2009), 7–17. For a wide-ranging and readable analysis of how and why we may become dependent on packed sensuality, see David Linden, *The Compass of Pleasure: How our Brains Make Fatty Foods,*

*Orgasm, Exercise, Marijuana, Generosity, Vodka, Learning, and Gambling Feel So Good* (New York: Viking, 2011).

4. Kessler, *The End of Overeating*, 35–62, 168.

5. Ibid., 138–68; Daniel Lord Smail, *On Deep History and the Brain* (Berkeley: University of California Press, 2008), 118–47.

6. Kenneth J. Meier, *The Politics of Sin: Drugs, Alcohol, and Public Policy* (Amonk, NY: M. E. Sharpe, 1994), 48, 65; Maureen Ogle, *Ambitious Brew: The Story of American Beer* (New York: Harcourt, 2003), chap. 3.

7. Cassandra Tate, *Cigarette Wars: The Triumph of the Little White Slaver* (New York: Oxford University Press, 1999); Marc Linder, *"Inherently Bad and Bad Only": A History of State-Level Regulation of Cigarettes and Smoking in the United States since the 1880s* (Iowa City: University of Iowa Faculty Books, 2012), http://ir.uiowa.edu/books/2/.

8. John McWilliams, *The Protectors: Harry J. Anslinger and the Federal Bureau of Narcotics, 1930–1962* (Newark: University of Delaware Press, 1990).

9. Robert N. Proctor, *Cancer Wars: How Politics Shapes What We Know and Don't Know about Cancer* (New York: Basic Books, 1995), 110–11; Proctor, *Golden Holocaust: Origins of the Cigarette Catastrophe and the Case for Abolition* (Berkeley: University of California Press, 2012); David Courtwright, *Forces of Habit: Drugs and the Making of the Modern World* (Cambridge, MA: Harvard University Press, 2001), chaps. 4 and 8; Philip Hilt, *Smoke Screen: The Truth Behind the Tobacco Cover-Up* (Reading, MA: Addison-Wesley, 1996), chaps. 1 and 2.

10. Peter Stearns, *Fat History: Bodies and Beauty in the Modern West* (New York: New York University Press, 1997), 22.

11. Hillel Schwartz, *Never Satisfied: A Cultural History of Diets, Fantasies, and Fat* (New York: Anchor Books, 1986), 5, 11, 83–105, and chap. 5; Andrew Smith, *Eating History* (New York: Columbia University Press, 2009), 113–23; Stearns, *Fat History*, 12–56, 105–12; John Coveney, *Food, Morals and Meaning: The Pleasure and Anxiety of Eating* (London: Routledge, 2000), xiii; Robert H. Lustig, *Fat Chance: Beating the Odds Against Sugar, Processed Food, Obesity, and Disease* (New York: Hudson Street Press, 2012).

12. Gary Cross and John Walton, *The Playful Crowd: Pleasure Places in the Twentieth Century* (New York: Columbia University Press, 2005), 97–100.

13. Bruce Bliven, "Coney Island for Battered Souls," *New Republic* 28 (1921): 374.

14. Good examples include Frank R. Leavis, *Mass Civilization and Minority Culture* (Cambridge: Minority Press, 1930); and Lewis Mumford, *Sticks and Stones: A Study of American Architecture and Civilization* (New York: Norton, 1924).

15. Cross and Walton, *Playful Crowd*, chap. 3.

16. Émile Durkheim, *Suicide: A Study in Sociology* (1897; Glencoe, IL: Free Press, 1951), 246-58; see also Durkheim, *Division of Labor in Society* (1893; Glencoe, IL: Free Press, 1964), 17, 353-73.

17. José Ortega Y Gasset, *The Revolt of the Masses* (1930; New York: Norton, 1957), 7-8 and chap. 3; Bernard Rosenberg and David M. White, eds., *Mass Culture: The Popular Arts in America* (New York: Free Press, 1957); Daniel Horowitz, *Consuming Pleasure: Intellectuals and Popular Culture in the Postwar World* (Philadelphia: University of Pennsylvania Press, 2012), chap. 2.

18. Aldous Huxley, "Pleasures," in *Aldous Huxley, Complete Essays: Vol. 1, 1920-25*, ed. Robert S. Baker and James Sexton (Chicago: Ivan Dee, 2000), 354-57; Robert S. Baker, *Brave New World: History, Science, and Dystopia* (Boston: Twayne, 1990), 10, 73-98; Aldous Huxley, *Brave New World* (1931; New York: Harper and Row, 1965), xvii-iii, 52, 78, 268, 284; Laura Frost, "Huxley's Feelies: The Cinema of Sensation in *Brave New World*," *Twentieth Century Literature* 52 (2006): 443-73; Laura Frost, *The Problem with Pleasure: Modernism and its Discontents* (New York: Columbia University Press, 2013).

19. James Rorty, *Where Life Is Better* (New York: Reynal & Hitchcock, 1936); Louis Adamic, *My America* (New York: Harper, 1938); Erskine Caldwell, *Some American People* (New York: McBride, 1935).

20. Gary Cross, *Time and Money: The Making of Modern Consumer Culture* (London: Routledge, 1993), chap. 5.

21. Jesse Steiner, *Americans at Play* (New York: Harper Brothers, 1933), chap. 3; President's Committee on Recent Social Trends, *Recent Social Trends in the United States*, II (New York: McGraw-Hill, 1933), 995; Jeff Wiltse, *Contested Waters: A Social History of Swimming Pools in America* (Chapel Hill: University of North Carolina Press, 2007).

22. "Introduction," President's Research Committee on Social Trends, *Recent Social Trends*, I, liii.

23. Kelly Brownell and Katherine Horgen, *Food Fight: The Inside Story of the Food Industry, America's Obesity Crisis and What We Can Do About It* (New York: McGraw-Hill, 2004), 12-27, 229-31. A critique of this approach can be found in

Sander Gilman, *Fat: A Cultural History of Obesity* (Cambridge, MA: Polity Press, 2008).

24. Jim Blascovich and Jeremy Bailenson, *Infinite Reality: Avatars, Eternal Life, New Worlds, and the Dawn of the Virtual Revolution* (New York: Harper-Collins, 2011), 176–85; E. Aboujaoude, L. M. Koran, N. Gamel, "Potential Markers for Problematic Internet Use: A Telephone Survey of 2,513 Adults," *CNS Spectrums*, 11 (2006): 924–30.

25. Blascovich and Bailenson, *Infinite Reality*, 191–255.

26. Simon Reynolds, *Retromania: Pop Culture's Addiction to its Own Past* (New York: Faber and Faber, 2012), 75; Shimon Edelman, *Happiness of Pursuit: What Neuroscience Can Teach Us about the Good Life* (New York: Basic Books, 2012).

27. Robert Putnam, *Bowling Alone: The Collapse and Revival of American Community* (New York: Simon and Shuster, 2001).

28. Arguing against the view that the Internet causes psychological and social problems are Katelyn Y. A. McKenna and John A. Bargh, "Plan 9 from Cyberspace: The Implications of the Internet for Personality and Social Psychology," *Personality and Social Psychology Review* 11 (2000): 57–95. Contrast the critique set forth by Brad Bushman and B. Gibson, "Violent Video Games Cause an Increase in Aggression Long after the Game Has Been Turned Off," *Social Psychological and Personality Science* 1 (2010): 168–74. And for a broader synthesis, James Brook and Iain A. Boal, *Resisting the Virtual Life* (San Francisco: City Lights, 1995).

29. David Shenk, *Data Smog: Surviving the Information Glut* (New York: Harper, 1997); Reynolds, *Retromania*, chap. 2; Nicholas Carr, *Shallows: What the Internet is Doing to Our Brains* (New York: Norton, 2010); Richard Foreman, "The Pancake People, or, 'The Gods are Pounding My Head,'" August 3, 2005, http://www.edge.org/3rd_culture/foreman05/foreman05_index.html, cited in Reynolds, *Retromania*, 73.

30. Russell Belk, *Collecting in a Consumer Society* (London: Routledge, 1995), 29–35; Reynolds, *Retromania*, chap. 3.

31. Gary Cross, *All-Consuming Century: Why Commercialism Won in Modern America* (New York: Columbia University Press, 2000), chaps. 1 and 8.

32. Smail, *On Deep History and the Brain*, 117; Kessler, *The End of Overeating*, 244–45, 184–88, 192, 206, 204.

33. Susan Linn, *The Case For Make-Believe: Saving Play in a Commercialized World* (New York: New Press, 2008).

34. Mark Griffiths and Mark Davies, "Does Video Game Addiction Exist?" in *Handbook of Computer Games Studies*, ed. Joost Raessens and Jeffrey Goldstein (Cambridge, MA: MIT Press, 2005).

35. Stephen Kline, Nick Dyer-Witheford, Greig De Peuter, *Digital Play: The Interaction of Technology, Culture, and Marketing* (Montreal: McGill-Queen's University Press, 2003), 84-108.

36. John Beck, *The Kids are Alright: How the Gamer Generation is Changing the Workplace* (Cambridge, MA: Harvard Business School Press, 2006); Steve Johnson, *Everything Bad Is Good for You: How Today's Popular Culture is Actually Making Us Smarter* (New York: Riverhead Books, 2005).

37. Edward Palmer, *Children in the Cradle of Television* (Lexington, MA: Lexington Books, 1987), 32-36; Michael Pertschuk, *Revolt against Regulation* (Berkeley: University of California Press, 1982) 12, 69-70.

38. Edelman, *Happiness of Pursuit*.

39. Wendy Parkins and Geoffrey Craig, *Slow Living* (Oxford: Berg, 2006), 52; James Gleick, *Faster: The Acceleration of Just About Everything* (New York: Pantheon, 1999), 86-87; Carl Honoré, *In Praise of Slow* (London: Orion, 2004), 4; Geoff Andrews, *The Slow Food Story: Politics and Pleasures* (Montreal: McGill-Queen's University Press, 2008); Carlo Petrini, *Slow Food Revolution: A New Culture for Dining & Living* (New York: Rizzoli, 2006).

40. Darrin McMahon, *Happiness: A History* (New York: Atlantic Monthly Press, 2006), 12-13, 199, 204, 241; Steven Pinker, *How the Mind Works* (New York: Norton, 1997), 389-93.

41. David Blanchflower and Andrew Oswald, "Well-Being over Time in Britain and the USA," *Journal of Public Economics* 88 (2004):1359-86; Richard Easterlin, "Will Raising the Incomes of All Increase the Happiness of All?" *Journal of Economic Behavior and Organization* 27 (1995): 35-48; and for a critique of this, Betsey Stevenson and Justin Wolfers, "Happiness Inequality in the United States," *Journal of Legal Studies* 37 (2008): 76 (for data).

# Index

kets "full line" of phonographs, 151; on music as mood adjustment, 154, 260; recording sessions, 143–45; rescues "fugitive" sound, 137; ponders clerical automation, 137; praised for immortalizing speech, 136; rejects Rachmaninoff, 152; reluctant to abandon acoustic recording, 253; short-sighted, 152–54; supplies "home entertainment," 142–43, 160–62; as trademark, 157; vs. Berliner, 139, 144; vs. Columbia Records, 140; vs. Mutoscope and Biograph Company, 195, 199; vs. Victor, 156–57; Vitascope, 196; wax cylinders, 137–45, 148

Edison's National Phonograph Company, 147–48

Eisenberg, Evan, 132

Elephantine Colossus, 222

empty calories, 105, 245

*enfleurage*, 27

epistemology of smell, 5

Eskimo Pie, 116

essential oils, 27

Faber, Joseph, 132–33

fan magazines, 202

Faraday, Michael, 171

fast-food restaurants, 247–48

FDA (Food and Drug Administration), negligence of, 278

feast days, 210

feasting, orgies of, 9

feature-length films, 201–2

fermentation, 22, 26–27, 117

Ferris wheels, 220, 233, 236

festivals, premodern, 209

fireworks, 216

Five Foot Shelf of Books, 55–56

flavor profiles, 16, 112

Fletcherism (ritualistic hyperchewing), 277

flue curing of tobacco, 78–82, 303n28

flutes, world's oldest, 5–6

Ford, Henry, diatribe against tobacco, 77

Foreman, Richard, 284

Fox Film, 200

"free choice" as fetish, 3, 272–74, 284, 304n38

frozen foods, 245

Fry, Joseph Storrs, 108

funneled fun, 16

gendering of sweet foods, 103, 110, 121

Genesee Pure Food Company, 102

Gibson Girls, 188

*Gilgamesh*, 25

Gillette's safety razor, 55

gin: addictive use of, 10; candy imitations, 106; consumed in near-epidemic quantities, 30; distillation of, 28

ginger ale, 125, 127

glass-blowing, 22–23, 40–41

glass jars, 39

Gleick, James, 264

gluttony and sloth, 90, 117, 272

Goo Goo Cluster, 113

Gorky, Maxim, 278

grain elevators, 97

grains: moral superiority of, 9; dominant in commoners' diet, 91

gramophone, 139

Grant, President Ulysses S., 78, 126

Great Atlantic and Pacific Tea Company (A&P), 245